The
Ultimate
Book of
Craft
Beer

This book is dedicated to all those who stand up for others in beer and in life.

To Ben, who never lets me down.

And to Kate, who always makes me laugh.

The Ultimate Book of Craft Beer

A compendium of the world's best brews

Melissa Cole

Hardie Grant

BOOKS

Contents

Introduction

Ultimate, it's a lot to live up to in a title isn't it? So, let's be honest here, it's not the bible of all things craft beer because you simply wouldn't be able to pick it up!

What I have set out to do is give you a solid tool kit to get the best out of your beer drinking exploration and, if you feel so inclined, you can share it with friends who would also like to know how to crack the code of craft brews.

From deciphering what you, or others, might like to drink based on what you already enjoy, to discovering beer styles you may never have dreamed existed, to how to cook or pair them with food, it's more of a lifestyle guide for enjoying beer.

It's not just about boozing all the time either, there's a selection of great alcohol-free beer offerings in here as well; seriously, great non-alcoholic beers exist and they are amazing for enjoying social situations when you're the evening's designated driver or moderating your alcohol intake or – dare I say it – calories.

Throughout, you'll find little snippets of information, nuggets of trivia and historic facts that will make you the envy of your friends and keep you warm at night as you hug your pub quiz prize...

And to become a black belt in beer, you'll need to know how it's made, what its foundations are, the raw ingredients that bring it to life. Once you've read this book, you'll understand the mixture of science and alchemy that creates the world's favourite alcoholic drink and how there really is something for everyone out there.

So, settle in, grab a beer of your choosing and put it in a glass, chug it from the can or swill it straight from the bottle because, the final thing I want to say to you before you embark on your beer journey, is this: I am not here to judge you. How you choose to spend your hard-earned money is up to you – drink what you like and how you like – I just want to share some of the things I love the most, with passion and care, in the hope that you might enjoy them too.

So here's my first piece of advice: Enjoy More

OK, that's not an edict to go and drink more, I just wanted to ask you to stop and smell the beer (and the roses if there happen to be any nearby!). The reason I say this is that all too often we don't take the time to appraise our beers before guzzling them down.

Now, I'm not saying that you not smelling your beer is necessarily a bad thing. Just finished mowing the lawn in the hot sun? Grabbing a frosty one and finishing the first third in one gulp is a thing of great beauty. Just finished a hard sports game or a run? Heck, grab that cold beer and reward yourself (in fact, science says you're doing a good thing, as beer is proven to have better recovery properties than water), but if you want to really get to know, and understand, what it is you're drinking, then you have to get your nose involved.

Why do I say that? Well, it's because the nose knows. It's your very first line of defence against nasty tastes and poorly made or kept products. It's a result of evolution, it's what tells us whether something is good or bad and, most likely, whether we'll like it or not.

As someone who judges beers the world over, I can frequently discard 30–50% of rounds of beer on smell alone. I can tell that there's something wrong and why just from the smell of it (not that I'm suggesting you will be able to... just yet!) and if you stop and smell your beer, using cautious little bunny sniffs first before inhaling deeply, then you'll really get to grips with what it is that you are about to try.

The other reason I advocate stopping to smell your beer is this: if you are having the very best night out with friends, a romantic evening with a loved one, or just a peaceful, contemplative moment with yourself, then if you stop and smell the beer, the next time you drink it, or perhaps even years from now, you may be gifted with the memory of how you felt in that moment. Smell is the most evocative of memory nudges and that, my friends, is worth its weight in golden ale.

My very first starting point when people are asking me for beers to try, is to ask them what other drinks they normally enjoy drinking and it has proven to be the most foolproof way of starting them on their beer journey. So here is a simplified

If you like this

Dry white wine

Medium white wine

Sweet white wine

Dry rose

Sweet rose

Light-bodied red wine

Medium-bodied red wine

Full-bodied red wine

Dry sparkling wine

Medium sparkling wine

Sweet sparkling wine

Natural wine/sherry

Gin and tonic

Dark spirits and coke

Straight golden and dark spirits

chart to help guide you or to use as a tool for helping others. One final tip is to initially try these beers at a similar temperature to the drinks you're used to, and make any changes you want as you go along.

Try this

A classic saison or grisette, a New Zealand-hopped pale ale or dry-hopped lager

Session IPA or classic Czech Pilsner

Juicy IPA or pastry sour

Fruited Berliner Weisse or Gose

Fruited wheat beer

US-style amber ale or UK-style mild

Extra special bitter (ESB) or dunkelweizen

Stout, porter or wee heavy

Helles or Kölsch

Crystal weizen or UK-style golden/summer ale

Fruited pale ale or faro

Lambic, gueuze or other mixed-fermentation beer

Saison, grisette or biere de garde, well-chilled

Dark lager, rauchbier or pastry stout

Beer aged in a barrel previously used for your usual spirit of choice

Raw Ingredients

The soul of any beer is its raw ingredients and every single one has a vital part to play.

In today's hop-obsessed beer market, you could be forgiven for thinking that nothing else matters, but just as it's often the unsung ingredients, like perfect seasoning, that makes all the difference to a fantastic plate of food, so it is with beers.

Brews that are carefully constructed for all the ingredients to work in harmony are often the best; it's a bit like a great band: just because one member will occasionally step forward for a solo doesn't mean the rest of the band isn't still there backing them up!

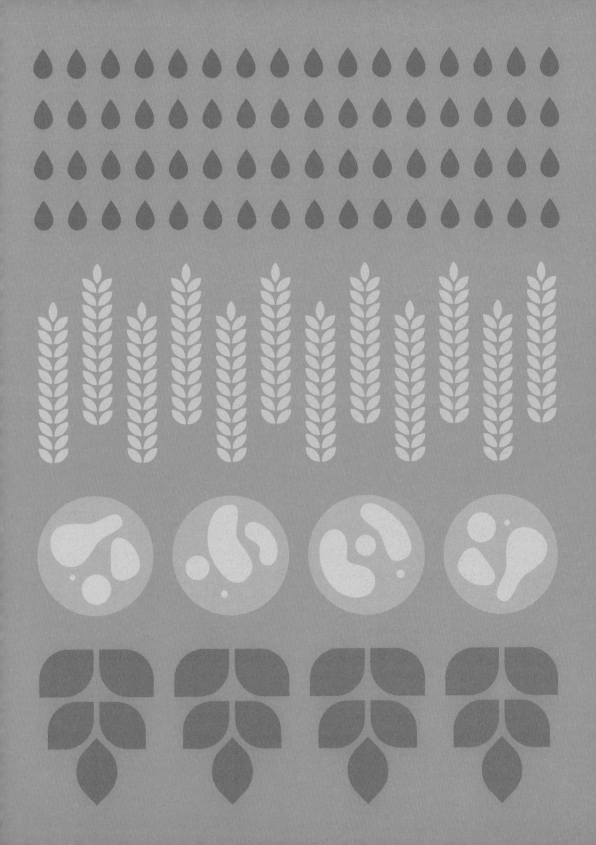

Grains

The structure on which every beer stands is its malt base, but what is malt exactly? To put it really simply, it's a grain, normally barley, that's been fooled into thinking it's springtime to start germination, and then stopped in its tracks by a careful drying and toasting process.

The level to which you toast the grains in the malting process gives you not only lots of accessible starches and active enzymes in the grains – the latter of which will convert the starches into sugars for the yeast to munch on and produce alcohol – but also accounts for colour and flavour in your beer.

The flavours (and colours) from malts can range from the lightest white bread, through wholemeal, to caramel and toffee, to raisin and milk chocolate to dark chocolate, through to the deepest espresso. These can then be mixed with a host of other malted and unmalted grains like wheat, oats, sorghum, rye and others to create as simple or as complex a flavour profile as you could wish for.

The very best malting barley is considered to come from maritime climates, and the UK is said by most brewers to have the best in the world (although, of course, that may also be a little bit of the good kind of national pride talking!).

Fact: Maris Otter is a species of barley grown in the UK that is famous the world over. The first species of barley that was deliberately bred for making beer, it is still prized for its flavour more than 50 years after being introduced.

Water

Water, water everywhere – but not for making beer... well, OK, that's not strictly true, but if you want to lay a bit of beer knowledge on people, then you can tell them that in the brewhouse, the water used for brewing is called 'liquor' and 'water' is used for washing things. I know, it's all a bit unnecessarily complicated, but that's brewers for you!

One of the reasons for differentiating is that brewers regularly treat the water they get, often from the mains, with what are known as 'brewer's salts'.

This is the process of treating the water with different minerals to mimic the natural water source of classic beer styles.

So, for example, one of the reasons that Burton-on-Trent became renowned for its IPAs is that its natural water source is very rich in gypsum, or calcium sulphate as it's scientifically known, which creates a more pronounced dry, clean bitterness.

Whereas London, historically renowned for its porters, is known to have a calcium carbonate-rich water supply, more suited to malt-forward beers and creating a rounder mouthfeel. Alternatively, you could class them as soft and hard waters respectively, but where's the fun nerd factor in that?

Hops

'A wicked and pernicious weed,' said someone at some point, although contrary to myth and legend, there is no evidence it was King Henry VI... but it's still an accurate description of the hop.

It is a climbing plant, which, if you look closely, can often be found running wild in hedgerows and even urban settings (near me in south-west London there's an alleyway with a rather large and luxuriant hop plant that's been left to grow untamed). However, its use in, and cultivation for, modern brewing now means that hops are used in nearly all but the tiniest percentage of the world's beer.

Hops, like grapes, develop a terroir. In case you were frightened to ask anyone what terroir means (like me when I first heard it) it's the effect that the hours of sunlight, the soil and the general climate has on a crop. In this case I'm talking about hops, but you'll more generally hear it spoken of in terms of wines.

Regardless of their original species, within a decade or so of having been planted in a different country, hops will develop characteristics that are so far away from their forebears that they are often renamed. For example, NZ Cascade has now been renamed Taiheke, but it originated from the Cascade hop that first came from a cross between a UK hop and a European hop. Every step of the way has been important in the development of that hop's characteristics, but the defining factor in what it smells and tastes like today is the terroir that it has been shaped by most recently.

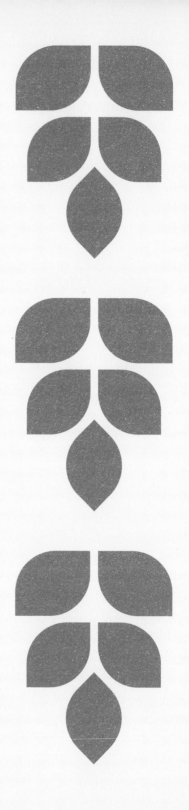

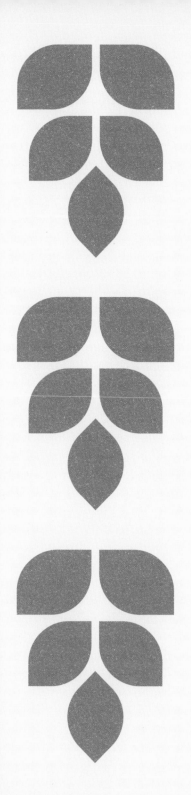

Anyway, my point is that hops have followed a similar path to that of 'old' and 'new world' wines a few decades ago, which brings us neatly back to that word 'terroir'.

Traditionally the UK has been known for its subtly bitter hops, with restrained aroma characteristics like tobacco, hedgerow fruit and mown grass. A lot of Central and Eastern European hops have similar characteristics to UK hops, but many of them have also been developed specifically for delicate styles like Pilsners and other lagers or ales, so you get more subtle and restrained notes like woody herbs, black pepper and fresh hay.

The USA is known for its more aggressively bitter hops, with big citrus, pine, rose and marijuana aromas and flavours. Australian hops tend towards the US versions with some fairly hefty bitterness, but with more subtle apricot, peach, lemony flavours and aromas and the odd foray into more floral lands.

Then there are New Zealand hops, which have some of the most interesting and complex flavours being developed anywhere (in my opinion). Their bitterness can be subtle to booming and aromas and flavours range from Sauvignon Blanc to lime zest, cherry blossom and beyond.

Did you know... Just as growers in 'old world' wine countries like France and Spain have started to develop lighter, more accessible and often more tropical flavours, so have 'old world' hop-growing countries like the UK, Germany the Czech Republic turned to 'new world' countries like the USA, Australia, New Zealand and Canada to ape their very popular hop characteristics.

Yeasts and Beasts

Yeast, that beautiful single-celled fungus that takes in sugars and expels alcohol and carbon dioxide, is the most critical ingredient in beer that was considered a magical element before science identified its presence.

Ancient civilisations believed that deities gifted the alcoholic element of their brews, the most famous evidence for which is the Hymn to Ninkasi, the Sumerian goddess of beer. The praise for this goddess was immortalised in a stone tablet in 1800 BC and also doubles up as a recipe for beer. And even the infamous Rheinheitsgebot, the German Purity Law, didn't recognise yeast as one of the core ingredients of beer when the first draft was written in Bavaria in the early 1500s. But now, thanks to technology, we understand more about yeast than we ever have before.

Yeasts in brewing are now so diverse that it's difficult to explain things simply, but here are the main types of the Saccharomyces family.

Lager Yeasts

The strain of yeast most commonly used for lagers is Saccharomyces pastorianus (named for Louis Pasteur who identified it as the most efficient yeast to make cold maturation beers). These diligent yeasts like to operate at cool temperatures over a week or so and then, quite literally, chill out for a few weeks, by which I mean stay in cold storage (this is where we actually get the word 'lager' from – it's derived from the German word to store).

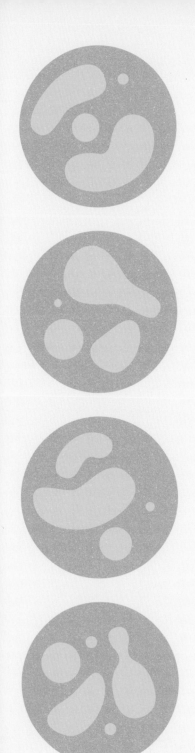

Ale Yeasts

Saccharomyces cerevisiae usually likes warmer temperatures and will ferment quickly, without the need to be held at very cold temperatures for a long time. These yeasts tend to create fruity or spicy flavours and don't eat all the complex sugars, so leave a little more of a rounded mouthfeel.

Beasts

There are also some other microscopic organisms that like to get in on the act. Brettanomyces is one of them. Meaning 'British fungus', it was first classified from British porters and is what is responsible for the 'funk' in certain beer. Much slower to act than other yeasts, it also tends to eat up all the sugars that its faster-acting cousin leaves behind, resulting in a bracing dryness and complex aromas ranging from high tropical notes to sweaty horse blankets (the latter not always terribly desirable!). You will find it in Lambic-style beers and some saisons and imperial stouts (although it's used more widely today by experimental brewers, this is where you will find it traditionally).

Finally, the bit that may freak you out a little, but please bear with me here! There can also be bacteria involved in making beers... WAIT, don't run away. The main strain that is used is also the stuff that makes yoghurt, which is called Lactobacillus and provides a lovely zingy sourness, the key characteristics of beer styles like Berliner Weisse or Gose. And there's also something called Pediococcus, which produces a stronger sourness. Finally, very occasionally, you'll find Acetobacter as well, but that is extremely rare and found only in very few beer styles due to its aggressive nature – after all, not a lot of people appreciate vinegary beer!

A Quick Guide to the Brewing Process

This is an incredibly simplified way to explain the brewing process, but to try and do it any other way would have to take into account the way that every single brewery in the world operates very slightly differently, and it turns out I don't have enough time, pages or will to live to do that!

Mashing in, which is adding crushed malted grains (and sometimes sugars or other fermentable materials) to a big pot with heated water, in order to create the right environment for the enzymes in the barley to convert the starches in the grain to sugars.

This can stay in one vessel, or be transferred to what's known as a 'lauter tun', which is a more shallow, wider vessel, that makes the next process faster.

Next process is sparging, which is sprinkling more hot water over the top the grains to push through a sweet liquid, known as the wort, which then gets transferred into another vessel, called the kettle.

In the kettle, that sweet wort gets boiled, partly to sanitise it and partly to help get the bittering compounds out of the hops, more hops may be added later for flavour and aroma purposes.

When all this is done, the wort is then transferred, via a heat exchanger to cool it, to the fermentation vessel, where the yeast is added and the magic happens. From there, post-fermentation, it may be matured for varying periods of time and then packaged.

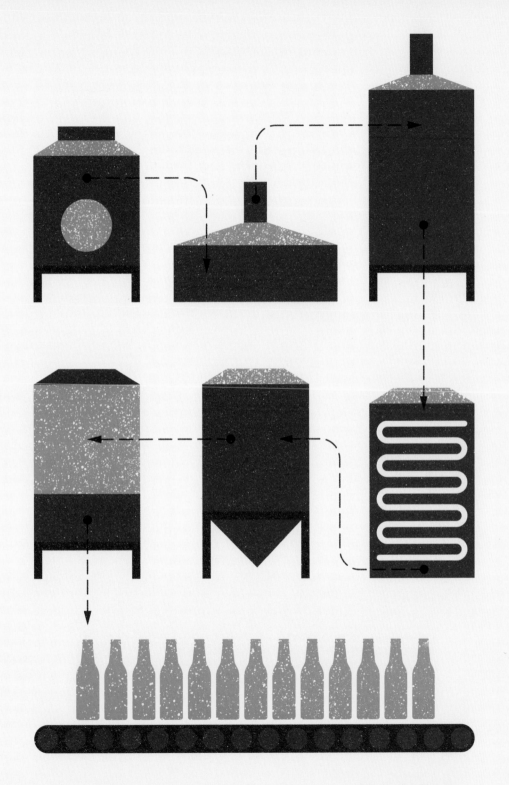

Beer and Food

Beer and food are the most natural of partners. From simple pairings like a ploughman's and a pint to more esoteric offerings like ale roast duck and barley wine, you can run the full gamut of tasty tie-ups.

But how do you start? Well, the first thing to remember is the old Army acronym K.I.S.S. – keep it simple, stupid.

Not that I'm calling you lovely people stupid, it's just that this is the most intelligent way to approach things, by starting simply and building confidence and knowledge of your palate and its preferred pairings from there.

The first bit of 'homework' for you is to get to know your beer. How intense is it in flavour? If it's a subtle, good-quality lager or a soft golden ale, then you want to keep your food very simple. Consider using just a few herbs and a bit of citrus, like a simple piece of white fish with a touch of thyme and a few slices of lemon, perhaps baked in a paper parcel with a splash of your chosen beer to steam it in, or maybe a goat's cheese salad.

If you are ramping up the flavours a bit with a Belgian-style wheat beer, which traditionally uses coriander seeds and orange peel in the recipe, then sushi or moules à la bière (moules marinière but replacing the wine with Belgian wheat beer) is a great way to go. As you get a bit more complex with the malt profile of your beer, such as subtle UK-style bitters or brown ales, you can introduce a more toasted character to your food.

A great Welsh rarebit/rabbit (a historic debate I'm not going into here!) or a fabulous rare roast beef sandwich with just a lick of horseradish and some peppery watercress could be your friend. Wandering more into the highly hopped territory of pale ales and IPAs, you need to go sweeter, so the classic pulled pork with tonnes of caramelised onions is a real winner (just make sure it's not too spicy unless you are a chilli masochist, as the CO_2 and, there is increasing evidence, the presence of a lot of hop bitterness in your beer will exacerbate the heat more and more as you eat and drink).

Sometimes, when the beer is really tropical in its hopping, using varieties like Mosaic or Citra, you can opt for vanilla ice cream with mango and passion fruit coulis drizzled over the top for a full fandango of fruity flavour. Then, as we get to the darker side, you must consider how much roasted bitterness your beer has. Deep-roasted stouts need to be balanced with sweetness, like milk chocolate; whereas bigger, boozier, sweeter offerings like barley wines and old ales can be used to balance out very salty foodstuffs, like big blue cheeses or a salt cod stew.

As more sour beer styles become available, you can use their acidity to balance out rich, indulgent flavours. Try slow-cooked beef cheeks with a dark sour beer that also has some earthiness from Brettanomyces yeast, like a Flanders-style red, or perhaps use a lighter, fruited Berliner Weisse as a match for a Christmas feast of partridge and pickled pears. And, of course, fruited sour beers are wonderful with indulgent chocolate desserts, helping to cut through their rich, cloying nature. However, that does also allow you to eat more pudding – a good or bad thing depending on your view of your waistline!

Throughout the book there are recipes using different beer styles and some suggested matches to go with them, so whether you opt to use those or head in your own direction, I'd love for you to let me know how you get on. Please do send me pictures and messages via social media of your beer and food fun (because that's what it should be, heaps of fun).

Twitter: @melissacole
Facebook & Instagram: @melissacolebeer
Web: www.melissacolebeer.com

Packaging and Dispense

You know that bit earlier (you read the intro... right?!) where I said 'put it in a glass, chug it from the can or swill it straight from the bottle'? Well, to use the annoying internet advert jargon, I just want to let you know 'one simple trick' that will allow you to get the most out of your beer.

Put it in a glass.

Yeah, yeah, I know, I'm not your mum but, honestly, most of your sensory access to the beer is removed when you don't. To put it more simply: if you can't smell your beer properly, you won't be able to taste your beer as well.

Also, sometimes, if you're drinking from a can in particular, the brewery might not have been that great at rinsing the cans after they've been through the packaging line, so you might get a taint of glue or sanitiser that makes you think less of the beer inside. But, don't let that stop you smashing back a tinnie at a BBQ on a hot day, or slugging a cold one down after you've just done a big run, because those are some of the best-tasting beers in the world.

A couple of additional pieces of advice:

- Don't buy green or clear bottled beer – they will have something called 'light strike' and it makes your beer smell and taste nasty. Basically, there are certain hop compounds that need protection from UVA/UVB light and neither green nor clear glass protects them from that (there is an exception for lambic-style beers, see that section on page 158).

- Cans will no longer give beer a 'metallic taint'; in fact, they are overall better for the environment as they are 100% recyclable, cheaper and lighter to transport, and they protect your beer from light and oxygen, which even the tightest crown cap on a bottle might not.

- Keep all beer you buy somewhere cool and dark, drink anything that is brewed to drink fresh as soon as you can and, preferably, keep it in the fridge.

Highball Glass

Shaker Glass

Wine Glass

Modern Sherry Glass

Brandy Balloon

Glassware

I could tell you that you should only drink certain beer styles from certain glasses, but as someone who consistently has to play a game of glassware Jenga® in their kitchen cupboard, I can't in all good conscience recommend it.

If you want to, then great – you do you, my fellow nerd. But there tend to be glasses already around your home that will do the trick. Got a highball glass? That's great for a Kölsch. A normal pint or shaker glass is great for lagers, summer ales and wheat beers. White wine glass? Crack out the saison, grisette and Berliner Weisse. Modern-style sherry glass? Cool, that's wonderful for really nosing and savouring mixed-fermentation beers. Big red wine glasses or brandy balloons are ace for big, boozy wood- and barrel-aged beers.

You're probably beginning to notice a pattern here: the fancier the drink, the fancier the glass. Just like it is with other drinks. Hence why I suggest you don't go completely mad on beer-specific glassware... well, not until you totally disappear down the beer rabbit hole!

Drink Analytical

When it comes to the role of smell in tasting, science still doesn't have it all worked out. But, what we do know is there are between 6–9 million olfactory neurons between the upper part of the nasal cavity and the back of the throat. These are divided into two systems; the ortho-nasal system built for the perception of odours during sniffing, and the retro-nasal system made for perception of flavour during eating and drinking.

The purpose of these two systems – it was quite recently discovered – is quite separate. The first is designed as an analytical tool for your brain to identify and catalogue smells, but the second, it would seem, is actually the opinion former as it converts aromas into flavours for the mind and imprints the memory of taste through scent – it's also the one most people can blame for being captivated by chocolate or hating Brussels sprouts!

This is because the retro-nasal system isn't hard-wired straight into the cognitive, or thinking, part of the brain. It takes a different, slightly meandering, route through the ancient seats of appetite, anger, fear and memories – or, more prosaically, the hippocampus, hypothalamus and amygdala.

This is most likely to have developed through evolution, from when tasting foods and liquids was a bit like playing Russian roulette and the human race learnt to inherently fear that which would make them ill or worse.

You remember the old 'tongue map' we learned in school, or you might have seen it in tastings? It shows the very front of the tongue as responsible for sweet, back from that is salty, then behind that sour and bitter – well, it's largely complete hooey!

This fallacy first appeared at the turn of the twentieth century, German research scientist D. P. Hanig did some research from which he concluded that there were areas more sensitive to various tastes. Over the years it was over-exaggerated and illustrated in this very linear way to explain our sensitivity to each taste known at the time: sweet, salty, bitter and sour.

There is also the 'fifth taste' and that's umami. Umami is the detection of savoury, and very appetite-inducing, glutamates and nucleotides. It was first scientifically identified in 1908 by Professor Kikunae Ikeda. There is undoubtedly an umami element to some beers, whether through the autolysis of yeast (dead yeast cells – not as gross as you'd think, but then they are used to make Marmite. I'm not sure where I'm going with this other then to say 'I hate it'!).

And, by the way, if you wonder why I'm not using 'flavour' and 'taste' interchangeably, it's because they aren't. Taste is mechanical, flavour is chemical and tied into genetic capabilities to detect those compounds.

Spotting Common Faults

These are the most common faults you'll tend to find in beer. There are many more – some of them extremely gross – but I'll try not to terrify you too much!

Diacetyl

A sweet butter popcorn aroma and slickness on the palate – think hard caramel candy – which is normally a brewing fault. It means they haven't taken time or care enough to go through the right fermentation or maturation processes. Sometimes, however, if you're drinking draught beer and the taste is like a hard candy that you've dropped in the dirt, that can be dirty lines in the venue. Maybe take a look around and see if everyone else is drinking packaged beer and follow suit!

Light Strike

Going back to the 'don't buy beer in clear or green glass' comment I made earlier (page 22), this is why. Americans call it 'skunky', personally I think it smells a lot more like tom-cat or fox pee, but then I live in London and we have foxes that are so well-fed that I think they should be re-classified as small urban wolves.

Oxidation

My most visceral reaction to oxidation is stale white pepper, damp dog or newspaper. It means that a beer that is supposed to have been drunk fresh is well past its sell-by date. However, it also happens in waves to beers that are suitable for ageing. Sadly, this is an inexact science but keep an eye on the social media of beer writers, beer geeks and brewers to see what they are drinking. If they are drinking something you have, and they say it's drinking well, you might want to crack yours open.

Phenols

TCP, Band-Aid® or peat or smoke in unacceptable levels generally means that someone has been inattentive in the brewhouse. Whether it's that they've been trying to get too much out of the ingredients in the sparging process (see 'A quick guide to the brewing process', page 18), they've not treated the water properly to get rid of the chlorine or there's an unwelcome wild yeast or bacterial infection.

Over-/Under-carbonation

Personally, I find over-carbonation a bigger problem than under-carbonation, but then I enjoy cask ale, which has the merest hint of bubbles – plus I've never had the desire to burp the alphabet. It could also be the sign of a wild yeast infection; if it smells funky too, then it probably is, so do let the brewery know.

Under-carbonation is usually poor packaging practice, but sometimes, if there's an additional fermentation happening in the bottle (called 'bottle conditioning' – why you'll sometimes find some yeast sediment at the bottom of certain beers), it's because someone has either forgotten to provide sugar for the yeast to eat and ferment or the yeast isn't healthy.

Sour/Acidic

Unless it is a style that is deliberately supposed to be sour, like a Berliner Weisse or a Gose, or mixed fermentation-style beer this is normally the result of a bacterial infection.

Burning/Bitey

My biggest bugbear in modern beer is breweries that rush beer out of maturation before the beer is properly finished. As a result you can get hop burn or yeast bite (due to there still being excess hop matter or yeast in suspension in the beer) and the beer also has a tendency to taste 'dirty'. There is no excuse for this; it's lazy and greedy. Time, temperature and patience are just some of the most important things in making a great beer, and there are way too many 'hype' brewers producing sub-standard brews and gaslighting drinkers into thinking it's how it should be. Also, if you're paying a big sum of money for these big-hopped beers that just aren't right, especially if they tell you to 'store it in the fridge for a few weeks and it'll be fine', I strongly suggest you spend your money elsewhere. They are taking the proverbial out of you. Oh, and cans shouldn't explode either. EVER.

OK, so I got that off my chest! Shall we move on to the good bit? Let's try some beers...

1

Lagers

What defines a beer as a lager? Well, 'lagering' is actually the term for an extended period of ageing at a very cold temperature.

However, we have come to see lager through the lens of styles like pilsner, which is the inspiration for most of the world's best-selling beers (albeit most of them are no longer properly lagered).

Lager comes in all shapes and forms, from Oktoberfest steins (for which you should travel to the festival in Munich to try them at their best) to dunkels, black lagers to commodity brands and far beyond.

I've tried to incorporate a few different styles in here, and you'll also find a bock, a strong lager, beer in the Red, Amber and Brown chapter (page 122).

So, I'd like to encourage you to spread your wings beyond the norm with this style of beer and to try beers that have a bit more (in some cases, a lot more) flavour than the well-known, rather watered-down brands.

Budweiser Budvar Reserve Extra

ABV
7.5%

Country of origin
Czech Republic

Try it if you like
Sweet bourbons

Great with
Roast duck

Also try
Birra Del Borgo
My Antonia,
7.5%, Italy

I have been lucky enough to visit the Budvar brewery on a number of occasions. It is a genuinely magical experience.

Still owned by the Czech government, it has been a global success, despite the long-running legal battles with beer behemoth AB InBev – I think it's fair to say, may not have been an entirely original thought to name its flagship beer Budweiser, but I'll leave it there. I can't afford the law suit.

Drinking the unfiltered, unpasteurised version of the normal beer, straight from the maturation tanks, was my lager epiphany. In fact, I'm fairly confident there might even be nail marks in the door frame where myself and fellow beer writer Pete Brown had to be dragged out of the place.

Banish any notions that this is some sort of 'special brew' – it has been aged for 200 days and it shows. Softly boozy with some complex pear and caramel notes, it settles into an orangey, grassy bitterness that is extremely pleasant to cut through fatty foods.

It's not one to neck with pals by the pint, but it would keep you warm in a cold Czech cellar, a bit like a brewmaster's down-filled coat.

Lost and Grounded Keller Pils

ABV
4.8%

Country of origin
UK

Try it if you like
Cava

Great with
A fresh pretzel

Also try
Big Rock Brewery
Pilsner
4.9%, Canada

When Alex Troncoso, formerly of Little Creatures in Australia and Camden Town Brewery in London, announced he was opening a brewery in Bristol with his partner Annie Clements, the beer world was waiting with pint glasses at dawn for the first brews, and they didn't disappoint.

The Keller Pils (cellar Pils) is a triumph because there are so few beers made in this style that package well, but Lost and Grounded have achieved it without filtering or fining, thus leaving a lot of the delicious bready character that comes with the German Pilsner style.

Being true to the style, they have also used typical German hops, meaning that there's a lot of cut-grass character and that lovely scrubby, herbaceous thing going on.

Schönramer Surtaler

ABV
3.4%

Country of origin
Germany

Try it if you like
Easy drinking

Great with
Grilled cod

Also try
Hönöbryggeriet
Lager
5%, Sweden

This is categorised as a *Schankbier*, which is an antiquated tax bracket for lower-alcohol lagers in Germany. But just because it no longer applies legally, doesn't mean it's not a great style to look out for when you are travelling around what is arguably the lager capital of the world...

Now in its eighth generation of family ownership, having been established in 1780 by farmer Franz Jacob Köllerer, who acquired the estate of the private brewery Schönram, it holds dear its Bavarian roots and ensures that the historic DNA runs through all of its brews.

The beer itself is light, refreshing perfection, walking a tightrope of balance between soft malt character and just a little hint of spice from the hops. A simple beer that harks back to simpler times – well, they probably weren't but with all the chaos in the world, it's nice to think that.

Croucher
Sulfur City Pilsner

ABV
5%

Country of origin
New Zealand

Try it if you like
Light, dry tropical wines

Great with
Simple grilled shellfish

Also try
Five Points Pils
4.8%, UK

What started off as a typical Czech-inspired Pils has, with the addition of those super-fruity New Zealand hops, morphed into something more akin to the love child of a Pilsner and a pale ale.

And it's a real achievement from this tiny brewery in an industrial estate in Rotorua, which has some of the most innovative solutions to brewing on a shoestring I've ever seen. However, when I visited the brewery the welcome was anything other than parsimonious, especially in its tap house in the town where they serve their beer alongside great food.

Zesty and tropical but still providing all the refreshing qualities you want from a lager, it is one that I genuinely hanker after on hot days.

The beer derives its name from the geothermal vents around the city, which give it astonishing geysers and warm water pools – and a rather unfortunate odour of rotten eggs. A bit overwhelming at first, it's quite easy to adapt to and worth a visit not just for the beer, but to learn more about the indigenous people's culture at the amazing Tamaki Māori Village.

Narcose Czech Amber Lager

ABV
4.7%

Country of origin
Brazil

Try it if you like
British bitters

Great with
Barbecued picanha steak

Also try
West Berkshire
Vienna Lager
4.8%, UK

Tucked away in Capão da Canoa – a town an hour and a half outside Porto Alegre in Brazil – is one of the most picturesque breweries you could imagine. From the taproom veranda you can survey the mountains, a lake and the sea, which is mere minutes away. One look and you'll never want to leave.

The Brazilian beer scene is one that is flourishing. With a huge influence from early German and Portuguese immigrants, and now from beer cultures globally, it has some of the most earnest and enlightened beer schools in the world.

The Diehl family, who own the brewery, are effectively my Brazilian família. They welcome me with open arms any time I visit, but that's not the only reason for their mention here. The beers that are coming out of the brewery are perfect for the weather, and have already won multiple awards in just four short years.

This is probably my favourite easy-drinker from the line up, with bread, digestive biscuit and the teeniest hint of toffee underneath an enticing floral aroma. It's simply perfect with grilled meat, which is pretty much the staple diet in Brazil!

Brooklyn Lager

ABV
5.2%

Country of origin
USA

Try it if you like
Whisky and soda

Great with
Roast chicken

Also try
Sierra Nevada
Oktoberfest
6%, USA

Brooklyn Lager was a genuine eye-opener the first time I had it, which was at a bar in New York – so pretty much fresh from source. While I'd had darker lagers from Germany and the Czech Republic, there was just something so distinctly different about it that I couldn't help but linger over it, taking it all in.

Brooklyn can only be described as a tremendous success story. It took on a site in, what was at the time, a deeply deprived neighbourhood in New York and put a lot of investment into the area. In recent years, the brewery has been extremely vocally supportive of the LGBTQ+ community.

It has also been hugely successful on a business level, having bought up some minority shares in other US craft brewers, like Funkwerks in Fort Collins, CO, and – one of my personal favourites – 21st Amendment in San Francisco, CA, as well as one with a rather dubious history: London Fields in the UK capital.

Now 24.5% owned by Kirin and partnering with Carlsberg to produce its beers in Europe, you can expect to see more of its flagship beer, with its rich brioche and marmalade edge and biting nettle hop bitterness.

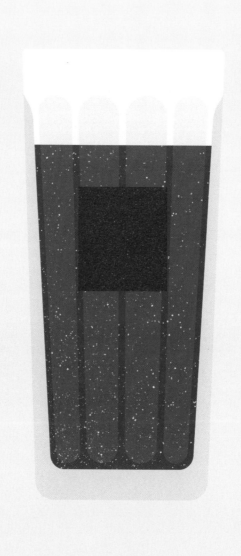

Augustiner-Bräu Dunkel

ABV
5.6%

Country of origin
Germany

Try it if you like
British bitters

Great with
Schnitzel

Also try
Calvors
Dunkel Lager
4.5%, UK

Dunkel is the classic, historic style of Munich and Augustiner is one of its finest proponents. Being one of the six official breweries that can produce Oktoberfest beers for the famous Munich piss up – sorry, celebration – Augustiner knows what it's doing when it comes to making lagers.

Brewing since 1328, it was run by monks for 500 years before the monastery went into state ownership in 1803 due to the secular reform happening in the country. It moved into private hands, and out of the crumbling monastery, in 1817 and, despite having a few different sites in that time, remains a Munich institution.

Light coffee and chocolate notes are augmented by a fresh cut grass and herbal hop character, but all in a really light, easy-drinking body. So don't be afraid of the colour, just enjoy the different experience of drinking a dark lager.

La Cumbre
Beer

ABV
4.7%

Country of origin
USA

Try it if you like
Prosecco

Great with
A hot sunny day

Also try
7 Peaks
La Dent Jaune
5.2%, Switzerland

Like everyone else, I have little rituals – silly daft things that make me happy – and one of those things involves this beer. It's always, always the first beer I drink when I get into the Great American Beer Festival, and brewmaster Jeff Erway knows to start pouring the minute he sees me coming!

You might be wondering why: why is this beer so special? After all, it's 'just a lager', right?

Wrong. There's a reason why brewers and beer writers like me love a great lager. First and foremost it's because to brew a truly great lager, without any flaws, is one of the hardest skills in brewing.

It's also because when you've been judging everything from imperial barrel-aged stouts to smoke beers for three days, there is nothing, and I mean nothing, finer than a brilliantly crafted lager like this.

Clean, fresh and full of flavour, it demands nothing more of you than to enjoy its zesty, zippy, lemony refreshment and incredibly clean fermentation and maturation – just like it should be.

Köstritzer Schwarzbier

ABV
4.8%

Country of origin
Germany

Try it if you like
Porters

Great with
Slow-cooked pork knuckle

Also try
Thornbridge
Lukas
4.2%, UK

Köstritzer is one of the oldest producers of *Schwarzbier* (black lager) in the world and it does it extremely well.

A true survivor of the Berlin Wall, it was one of the few German breweries that managed to keep up exports from East Germany during the great divide with the West.

Now owned by German brewing giant Bitburger, it still continues to turn out the most moreish of all the black lagers (in my opinion), with dried fruit and coffee notes and a hint of fresh tobacco. It is definitely one to try if you think that dark beer can't be light and refreshing.

Coedo Kyara

ABV
5.5%

Country of origin
Japan

Try it if you like
Orange wine

Great with
Garlic prawns

Also try
Lacons
Steam Lager
4.6%, UK

The Coedo Brewery has a fascinating history. It was founded on the principles of local sourcing and organic farming. However, when the owners realised they weren't able to find anywhere to malt the local barley, they looked to a different crop – sweet potatoes – and a new Japanese beer style was born.

However, Kyara is brewed in a more traditional fashion and the name denotes the Japanese word for a deep golden brown colour tinged with red.

Using the gloriously vinous and tropical Nelson Sauvin hop, the beer itself is also complex in body. It uses six different malts to great effect – each one building to provide a light orangey, rye-bread undertone that never dominates or feels cloying. It's a beer that's won multiple awards and, honestly, once you try it, you'll understand why.

Pilsner Urquell

ABV
4.4%

Country of origin
Czech Republic

Try it if you like
Peroni

Great with
Another one!

Also try
Pillars Brewery
Pilsner
4%, UK

You simply can't write a section on lager and leave out Pilsner Urquell, the brewery that spawned a thousand imitators and gave birth to the world's favourite beer style (albeit most of them are now pale imitations of where it all started).

The story of how this beer was created is no doubt somewhat sanitised and far from the reality, but the rough gist of it all is that the town of Pilsen was so disgusted with the beer that was being made there that in 1838 they poured it down the drain, drummed the brewer out of town and hired an angry young man called Josef Groll who had knowledge of English pale malting techniques and German lager yeast to come and brew at the brewery that was being built. In October 1842 the first beer was poured and everyone was delighted. It gained a reputation and the rest, as they say, is history.

Although I mentioned pale malt you'll note that PU (as it's affectionately called) is actually a golden hue. That's because they boil part of the wort in separate vessels to get the distinctive caramelisation in the beer. Add to that the local Saaz hops and a five-week lagering time and you have simple drinking perfection.

Birrificio Italiano Tipopils

ABV
5.2%

Country of origin
Italy

Try it if you like
Prosecco

Great with
Parma ham

Also try
Ninkasi Pilsner
Cold Fermented Lager
4.7%, USA

If you say 'Tipopils' to any beer aficionado, you will most likely elicit a little sigh as they recall happy memories sipping on this lager perfection.

It's a big thing to say that a beer is perfect, but this one really is as close as one can get to easy-drinking perfection without being dumbed down. Owner Agostino Arioli is also a perfect drinking partner: wryly funny, humble but incredibly passionate about his beers, and with a sudden smile that can light up a room.

Agostino's background is in science, which is not unusual for brewers or brewery owners, and it's that sense of precision and technical excellence that makes his beers, and his reputation, so very impeccable. Actually, I'll be really honest, he's known as one of the most punctilious nerds in the business, who records every variable going, which reflects in the quality of his beer.

The beer itself is unfiltered and unpasteurised and can sometimes look a little hazy, but is none the worse for it. It has a soft peachy aroma that overlays the bread dough aroma from the malt and is as elegant as Milan Fashion Week.

Crooked Stave
Von Pilsner

ABV
5%

Country of origin
USA

Try it if you like
Refreshment

Great with
Mushroom pizza

Also try
Berliner Pilsner
5%, Germany

When you say the name 'Crooked Stave', most people will immediately launch into raptures about founder Chad Yakobson's wild fermented beers, and rightly so, but this is the beer that you are more likely to see him and the brewers crushing at the end of a tough shift.

Based in one of my favourite places in the world, Denver, CO, Chad's laidback manner coupled with a deep scientific knowledge of brewing make him one of my favourite 'beer nerds' to chat to at any given time.

Von Pilsner is an unfiltered *keller* (cellar) beer that has the delicious white bread middle that marks out a great German-style Pils, but still has that bright, fresh, clean and absolutely smashable quality about it, which is what makes it a great first drink (or two) while you're deciding how complicated you'd like to get with the rest of his magnificent brews.

What Is a Stave?

A stave is one of the pieces of wood that is used to make a wooden barrel and sometimes you get a crooked one that you need to replace, which is kind of the ethos of the brewery – a bit wonky!

Interestingly, if you look closely at a wooden barrel, you'll notice that not all the staves are the same width. This is where the skill of the cooper, a master barrel maker, comes into play, shaping them to all fit perfectly.

Moon Dog
Beer Can

ABV
4.2%

Country of origin
Australia

Try it if you like
Corona (the beer that is!)

Great with
A hot sweaty gig

Also try
Croucher
New Zealand Pilsner
5%, New Zealand

There are times when a beer is so very perfect for the situation you're in that you don't even bother looking at the bar, and this is my abiding memory of trying Moon Dog for the first time.

I was in Australia and we had just finished judging duties, so it was time to let our hair down a bit and go dancing. In the hot bar we were in, busting our moves on the dance floor (God, I hope there's no video evidence), can after can of this beer was passed along the line so we could carry on enjoying ourselves with a refreshing beverage in hand, but without letting our inner nerd get in the way.

It works because, even though there is a negligible malt profile, it is packed full of citrus hops. It's designed to be a great version of those somewhat dubious 'tropical' lagers that you need to stick lime in to make palatable. (Sorry, I do try not to be a beer snob, but I'm afraid that's my line in the sand!)

Marble American Pils

ABV
4.2%

Country of origin
UK

Try it if you like
Budweiser

Great with
BBQing in the sun

Also try
1936 Biere
Lager
4.7%, Switzerland

Not only has Marble been at the forefront of the UK craft brewing movement for decades, it has also seen some of the best-known names in the industry as well. From the founders of Wander Beyond, Cloudwater Brew Co. and Blackjack Brewery, to pillars of the craft community at Thornbridge, Vocation and Salt Beer Factory – it's been quite the academy for beer alumni.

In its latest incarnation, the brewery has moved slightly away from its roots in central Manchester, having been born in the iconic Marble Arch pub, to a purpose-built site in Salford, where head brewer Joe Ince is bringing the brewery once more to the fore of the craft movement.

Son of founder Jan Rogers, Ince hasn't just inherited the brewery by birth, he's worked across other leading lights of the craft brewing world in the UK, like Magic Rock, and is keeping the slightly anarchic energy of it all alive and well.

It was hard for me to select just one beer from the Marble stable. It's a brewery that I rarely dislike a beer from, but this Pils just shaded it over the brewery's flagship Pint, because I believe the latter is best served from cask, and it would seem unfair to tantalise international readers with a format that's out of their reach! But, suffice to say, if you ever find yourself in Manchester, I can't recommend the following strongly enough: step off your train or plane, head to the Marble Arch, order a pint of cask Pint, drink quickly with it barely touching the sides, order another one and sit back and admire the beauty of the pub surroundings as you do.

But, back to the Pils. Refreshing hits of pine, lime peel and a teeny hint of brioche, ending on an orange- and rose-blossom high. It's the best of simple drinks and one I enjoy immensely.

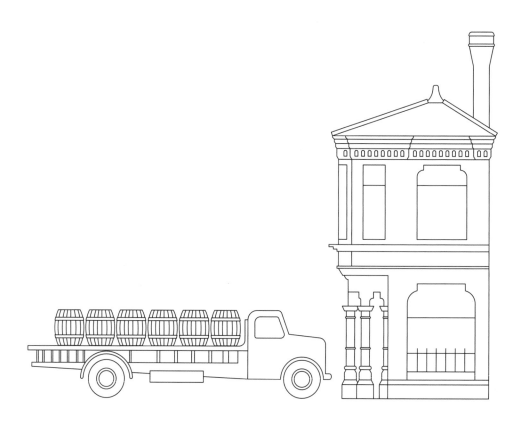

Rooster's Brewing Co.

ABV
4.8%

Country of origin
UK

Try it if you like
American Pale Ales

Great with
Cheeseburgers

Also try
Gaffel Kolsch
4.8%, Germany

Rooster's Brewing Co. – via my now mother-in-law – gifted me my epiphany beer, Rooster's Cream, which is now sadly only a seasonal; but it's a testament to how enduring a classic it is that it's still brewed (I say this as someone marking over two decades since this beer entered my life – god, that's depressing to write!).

Based in the beautiful Yorkshire town of Harrogate, the brewery may have changed hands from founder Sean Franklin to brothers Tom and Ol Fozzard, but the ethos stays the same: be proudly Yorkshire but welcome influences from all over the world.

Usually better known for their US-style beers – they make absolutely smashable pale ales and IPAs – they also collaborate with local companies like Taylor's of Harrogate, known for its high-quality teas and coffees, to make beers like a rose lemonade sour and green tea IPA, High Tea, which I am proud to have also collaborated on.

The Pilsnear, like all of Rooster's beers, is made with precision to be very slightly off-centre of the market. It isn't made with a lager yeast, but it is fermented and matured properly at cold temperatures, which, for me, is the most important part of its character.

Fresh and clean with lots of the delicious herbal, grassy, spicy, noble character from the traditional German hops used, it also has an overtone of something floral – a spicy marigold note, perhaps – that is so very light. It zips around like a busy firefly, never quite identifiable but providing a light note to the beer that makes it lip-smackingly delicious.

The brewery also has a lovely tap, that I strongly recommend making a visit to if you ever have the chance. It might be the wrong side of the Pennines according to my husband's family, but I always enjoy a warm welcome there and so will you.

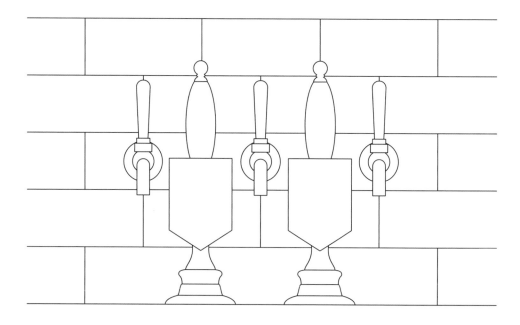

Früh Kölsch

ABV
4.8%

Country of origin
Germany

Try it if you like
Prosecco

Great with
People watching

Also try
Howling Hops
Das Köolsch
4.6%, UK

If you'd like to see an ignominious scuffle between beer nerds, just ask them which is their favourite Kölsch – Gaffel or Früh – and there's a distinct possibility it'll come to blows.

OK, so I'm exaggerating a bit, but for such a delicate beer style it does seriously raise some very strong feelings in people. I'm afraid I come down on the side of Früh.

Maybe this is because it was the first Kölsch I was aware of drinking. My late friend Glenn Payne got me one and, halfway through it, he asked me what I thought. He never asked idle questions when it came to beer, so I stopped and assessed it and realised I was drinking something a little more special than 'just a lager'. (Forgive me, I was younger and less educated then!)

Früh is a delicate delight, particularly when drunk fresh at source in Cologne. It's a beer I always get a delicate hint of lychee from, which I assume comes from its warmer fermentation. However, it's that cold lagering period that really gives it that crisp, lightly bready, totally palate-pleasing finish, which makes you reach for tiny glass after tiny glass of it.

Beery Tempura

- 1.25 kg (2 lb 13 oz) fish
  or 750 g (1 lb 10 oz)
  seafood or vegetables
  of choice, such as
  carrots, cut into
  2 cm (¾ in) batons
- sea salt
- groundnut (peanut)
  or grapeseed oil,
  for frying

For the tempura batter

- ice cubes
- 200 g (7 oz/1⅓ cups)
  plain (all purpose)
  flour
- 200 g (7 oz/1⅔ cups)
  cornflour (cornstarch)
- 1 medium egg
- 200 ml (7 fl oz/scant 1 cup)
  good-quality lager,
  put in the freezer
  until just about
  to freeze

I've not specified what you use the batter for here, because you can make your tempura with finely cut vegetables, thin strips of fish, whole prawns (shrimp) or scallops. Whatever you choose, make sure that you have two metal bowls that fit inside each other and plenty of ice to hand.

Prep the fish, seafood or vegetables and season lightly with sea salt. Place a wire mesh rack over a baking tray in a just warm oven and heat the oil in a deep fryer (this is safest) or a wok to 170°C (340°F).

To make the batter, first start by placing a layer of ice in the first, larger metal bowl (see intro above) and cover with a layer of paper towel.

Put the flours in the second, smaller metal bowl and mix well. Place the flour bowl inside the ice bowl.

Mix in the egg and beer lightly with chopsticks. Don't overmix – there should be lumps.

Dip your seasoned fish, seafood or vegetables in the batter and fry, without overloading the fryer, until light gold and crispy. Season lightly just before serving.

Beerita Sunrise

- 3 parts cold fresh lemonade
- 3 parts cold lager
- 3 parts tequila blanco
- 1 part grenadine
  cocktail umbrella,
  to garnish (because
  it's silly and fun!)

As this is a great party cocktail, I've given the recipe in parts – so you can make as much or as little as you wish.

Pour the lemonade, lager and tequila into a jug or large glass and stir gently. Pour into tall glasses.

Very gently pour the grenadine down the inside of each glass. One of those fancy, swirly bartender's spoons is good for this. It will swirl slightly with the bubbles but will mostly sink to the bottom, giving that sunrise effect.

Serve with the umbrella and drink while imagining you're on a beach in Mexico.

2

Wits, Wheats, and Weizens

Wheat beers can cover such a range of flavours it seems almost rude to plonk them all into the same category – I could wax lyrical about them all day.

Often you'll hear these beers categorised as Belgian or German in style, with the more recent addition of American-style, too. To be honest, the American-style wheat beer is often fairly unremarkable in that it doesn't have any huge yeast character or additional ingredients, but it is deliciously thirst-quenching – Goose Island 312 being a great example.

In contrast, Belgian wheat beers, while generally delicate, traditionally use the addition of coriander seeds and orange peel notes, and German wheat beers have everything from banana to clove to bubble gum in their make-up, depending on the strain of yeast used to make them.

Clear or cloudy, poured with a billowing head or a more subdued foam, however the beer presents, the key criterion is that 30–70 per cent of the grain bill* should be made up of malted or unmalted wheat to give it good body.

*The grain part of the recipe.

3 Floyds Gumballhead

ABV
5.6%

Country of origin
USA

Try it if you like
Starburst®

Great with
Kielbasa with
all the trimmings

Also try
Minoh W-IPA
9%, Japan

So, there aren't really many beers out there that have a comic strip made from the character on the labels. There are a few, but I am fairly sure this was the first.

Gumballhead the Cat, a 'bastard feline' that has been around since the early 90s, is illustrated by Rob Syers. Based on an ex-girlfriend's cat, Syers made Gumballhead into a character that drinks, smokes, stabs and crashes planes – which is, if I'm being honest with you, nothing I wouldn't expect from a cat that has apparently developed opposable thumbs.

In an interview in 2014, Syers claimed that when 3 Floyds put the shout out for art for this wheat beer, he went to see them and woke up with his legs tied to a keg, sinking into the spent grain of a drained lauter tun. He fought his way out of it, flattening folks in the brewery as he dashed to co-owner Nick Floyd's office, who then said "Give the kid some beer and take him home".

And the relationship was formed.

Knowing the guys at 3 Floyds, I'm not willing to say either way whether this is true or not, but given that I've been found in the afternoon drinking pints and shots with them, and toasting 'to evil' with every dram, I'm not sure I'm in a great position to judge them either.

It was a brew ahead of its time, combining spicy wheat beer characteristics with heavy American hopping – a forerunner to some of the white and Belgian IPAs you see on and off now as trends wax and wane. It is incredibly refreshing, but is to be respected for its ABV because it will slide down easier than a cartoon artist with his legs tied to a keg in spent grain.

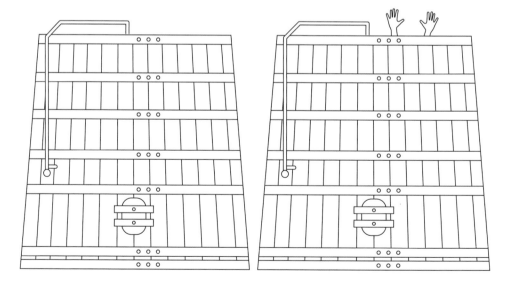

Hitachino Nest White Ale

ABV
5.5%

Country of origin
Japan

Try it if you like
Vodka and orange

Great with
Katsu curry

Also try
Moo Brew
Hefeweizen
5%, Australia

Japan's craft beer scene is constantly changing. Previously it was more obsessed with imports than it was with its own breweries, but that has slowly changed over the years and Hitachino has definitely been at the forefront of that.

I also have tremendous respect for how they fed and watered hundreds of people in the neighbourhood around them after the enormous tsunami and earthquake that hit the country in 2011, despite so much of their own brewery being affected.

Anyway, back to the beer. This ale has more spice than you'd normally expect from a Belgian-style beer, with nutmeg joining the traditional coriander (cilantro) and orange juice in addition to the expected peel, but all of it is done subtly, elevating this to more than just simple refreshment. It's a surprise with every sip.

Queer Brewing Flowers

ABV
3.5%

Country of origin
UK

Try it if you like
Medium sweet white wIne

Great with
Paella

Also try
Blanche de Namur
4.5%, Belgium

Kicking arse, taking names and breaking down barriers is something I admire in anyone, and I'm proud to call the founder of Queer Brewing, Lily Waite, my friend.

The first transwoman beer brand owner in the UK, and staunch advocate for diversity and LGBTQ+ rights, she continues to amaze me with her creativity and energy, and annoying ability to be excellent at most things (she's a tremendously accomplished artist and ceramicist as well).

Through partnerships with other breweries, she has raised a lot of funds for good causes in the queer space, and has worked hard to create a core range of beers for her own brand, which at the time of writing include a Pils called Tiny Dots, this witbier and a pale ale called Existence as a Radical Act, but I've no doubt there will be more before I've even finished proofreading this book, let alone by the time it goes to print.

The beer scene suffers from a surfeit of white beer bros, many of whom do nothing to help open doors for anyone who doesn't look like them. So the fact that Lily's brand is so boldly challenging this is something I can't see enough of; in fact, just recently, Cloudwater Brewing Co has helped not only Lily, but two black-owned and one Indian-owned brewery hit the shelves of a major supermarket via a mixed four pack. Long may this continue.

Flowers ticks all the boxes a wheat beer should: redolent of orange peel and coriander seeds, pops of herbal citrus hops and an ABV that means tucking away a few of these won't do too much damage in the morning.

Personally, I'd prefer a six pack of this over some blooms destined to die in a vase any day.

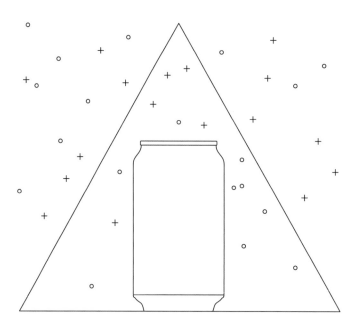

Maui Brewing Pineapple Mana

ABV
5.5%

Country of origin
USA

Try it if you like
Pineapple daiquiri

Great with
Poke bowl

Also try
21st Amendment Hell
or High Watermelon Wheat
4.9%, USA

After 16 years in the business that he founded with then wife Melanie, Garrett Marrero shows no sign of slowing down any time soon.

Founding the island's then only, and still the largest, brewery wasn't by any means easy, as Garrett might have been an enthusaistic imbiber, but he wasn't a brewer. In fact, he had a background in investment consulting, but that has stood him in good stead while building and growing the business that he loves – and which is now part of the fabric of the island.

Marerro has a deep commitment to the environment – the brewery garners much of its power from its solar panels – and a connection to the island's artistic and creative community. He is excellent company and a leading light for change – he's also well worth following on social media for his incredibly cute rescue dogs.

Full of fruity pineapple flavours, slightly acidic but only a touch, and rounded out by the wheat used in the brew to add a pleasing body. It's a bit like drinking sunshine in a can. You almost feel like you should have a lei around your neck and your toes in the sand as you sip it.

What Is Curaçao?

Curaçao is so much more readily associated with those terrifying blue drinks that you see on the shelf at the liquor store or in bars, but it's actually the name of the island from which the Laraha citrus fruit hails, and is the lesser known colonial story central to a globally popular beer style.

Curaçao is a constituent country of the Kingdom of the Netherlands, a Lesser Antilles island country in the southern Caribbean Sea, which was originally 'discovered' (in the Western way of looking at things that is, I'm fairly confident that the people who already lived there were pretty familiar with it) by Spanish explorers in the early 1500s, who also brought with them some Seville oranges.

Although the orange trees failed to produce edible fruit, upon being abandoned, they eventually changed by dint of terroir into a different species altogether, now known as the Laraha. This produces a very fibrous and unpleasant flesh but an extremely aromatic and appealing peel, which is dried and used in both brewing and the production of curaçao liqueur.

While it remains unknown who first produced the very first liqueur from the peel, through the Dutch West Indies Company (the equally rapacious Dutch equivalent to the British East India Company) supplies of the oils from the fruit made their way back to shareholders The Bols Distillery.

And the drinks company certainly made it famous, possibly because of the startling appearance of the liquid in the bottle. Lucas Bols, who ran the company from the late 1600s to the early 1700s, liked to add an air of mystery and flair to all his creations – hence the bright blue colouring added to the drink.

But how did it become synonymous with Belgian-style brewing? Well, it comes as a bit of a surprise to a lot of people that Belgium is actually a relatively new country, only formed as it's now known in 1830. It can sometimes be called more of an argument than a country, with fairly deep divides that run across the Flanders region in the north, which

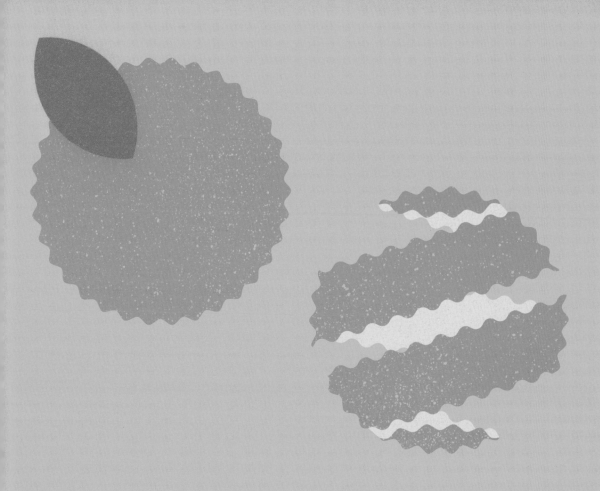

is predominantly Dutch speaking, and Wallonia in the south, which is predominantly French-speaking, and Brussels slap bang in the middle.

So, it's no real surprise really that there is a colonial hangover, so to speak, in the use of the spices – on which so much of the Dutch empire's wealth rested – in beers that hail from Belgium.

Of course, now the ability of people to get these sorts of ingredients via the internet at the mere press of a few buttons means that there are all sorts of different adaptations of this style knocking around. But the classics remain just that, and the versions that are true to the form have an almost unbeatable drinkability that seems to reach further than just the average beer audience, as Molson Coors Blue Moon has proven.

Fluffiest Falafel

Serves 4

- 300 g ($10\frac{1}{2}$ oz/$1\frac{2}{3}$ cups)
  dried chickpeas
  (garbanzos), rinsed
- 330 ml ($11\frac{1}{4}$ fl oz/$1\frac{1}{3}$ cups)
  Belgian wheat beer
- 30 g (1 oz) each of
  fresh mint, parsley
  and coriander (cilantro)
  leaves, roughly chopped
- 5 good-sized spring onions
  (scallions), white and pale
  green parts only, roughly
  chopped (reserve the dark
  green parts to garnish)
- 4 large garlic cloves,
  pounded to a paste
  with a bit of salt
- $1\frac{1}{2}$ teaspoons ground cumin
- 1 teaspoon fine sea salt,
  plus extra for seasoning
- $\frac{1}{2}$ teaspoon harissa
- enough groundnut, grapeseed
  or other neutral oil to
  cover 2 cm ($\frac{3}{4}$ in) of the
  base of a large frying
  pan (skillet)

To serve:

- lemon wedges (optional)
- good-quality extra-virgin
  olive oil (optional)
- toasted sesame seeds

There's no doubt that falafel has a bad name. Poorly done, it has a roof-of-the-mouth stickability that is second only to peanut butter but, done well, it's a crispy, fluffy ball of joy. I experimented with five types of beer versus water and the Belgian came out on top with everyone, offering bright citrus notes from its coriander seed and orange peel.

This recipe is a combination of J. Kenji López-Alt's recipe and that of a chef from my local Lebanese restaurant, plus my own beer twist. You can put these in a flatbread with some salad and dips or just add a blob of natural yoghurt, a squeeze of lemon juice, a sprinkle of toasted sesame seeds and a drizzle of extra-virgin olive oil and serve them as part of a meze or as a starter on their own. You can even pop them into a tagine right at the end if you like. If you check that your beer is vegan-friendly, then so is this recipe.

At least 12 hours before you want to cook your falafel, put the rinsed chickpeas in a bowl big enough to allow them to swell to three times their original size. Pour over the wheat beer and top up with enough cold water to cover by at least 5 cm (2 in). Refrigerate.

Drain the chickpeas and shake off as much liquid as possible, blot with paper towels and leave to air dry, spread on a baking sheet for around 10 minutes.

Put the dried chickpeas in a food processor and add all the other ingredients apart from the oil. Pulse until the mixture resembles fine breadcrumbs and just about holds together with a gentle squeeze. Refrigerate for 20 minutes.

Form the falafel mixture into golf ball-sized balls, and place on a lightly oiled plate.

Warm the oil in a large, deep frying pan (skillet) over medium-high heat. When ready to cook, gently lower the first falafel into the hot oil with a fork. Check it's lightly fizzing, not spattering, and that the oil reaches about a third of the way up (this should rise to halfway when more falafel are added to the pan).

Continue to add as many falafel as will fit in pan a 2–3 centimetres (1 in) apart, turn the heat to high for 15 seconds and lower back to medium-high again. Cook for about 4 minutes on each side, until golden brown. Repeat, if necessary, keeping the first batch covered in a warm oven.

When cooked, drain the falafels on paper towel, sprinkle with some lemon juice and a splash of good-quality olive oil, if you like, and some salt. Garnish with green spring onion tops and sesame seeds and enjoy warm (not hot).

NOTE: You can freeze the cooked falafel for up to 3 months, reheat them for 15–20 minutes in a preheated oven at 180°C (350°F/Gas 6).

Orange Beer Ice Cream

Makes about 1.5 litres
(52 fl oz) ice cream

- 5 egg yolks
- 140 g (5 oz/⅔ cup) caster
  (superfine) sugar or
  invertase (whichever
  is cheaper!)
- 400 ml (13 fl oz/1½ cups)
  full-fat (whole) milk
  (preferably Jersey, which
  has extra fat content that
  will make up for the beer
  not having any)
- 75 ml (2½ fl oz/⅓ cup)
  Belgian-style wheat beer
- 300 ml (10 fl oz/1¼ cups)
  double (heavy) cream
- zest of 2 oranges
  (preferably blood
  orange, if in season)
- 4 tablespoons orange
  juice (partially frozen)

Beer ice cream? Why not? The beer brings a depth and freshness to it that really works and it's a delicious accompaniment to a chocolate torte or warm brownie.

Whisk together the yolks and sugar in a bowl until they're properly amalgamated and very pale.

Gently heat the milk, beer and cream in a saucepan until it starts to bubble. Gradually whisk into the egg yolk mixture. Do not stop whisking or you'll get lumps!

Pour the mixture back into the saucepan and heat really gently. Stir until thickened (I find one of those silicone spatulas is best for this). DO NOT STOP STIRRING.

When you can draw a firm line in the mixture using the back of the spatula you are done.

Stir in the orange zest and partially frozen orange juice, whisk briskly, leave to cool to room temperature, then place in the fridge to get very cold.

When cold and firm, pass the ice cream through a fine sieve, in case you scrambled some egg. Churn in an ice-cream maker until set. Store in an airtight container and keep in the freezer until ready to serve. Allow to defrost for a few minutes before scooping!

A Sly Gin

Makes enough for 2 of those
big Spanish goldfish bowls
(or big red wine glasses)

- ice cubes
- 50 ml (1¾ fl oz/¼ cup simple
  syrup infused with ginger
  and lemongrass*
- 70 ml (2½ fl oz/⅓ cup)
  gin (I use Sly Gin Lemon
  Verbena, but dry London
  is good too)
- ½ lemon, plus 2 pieces
  of peel for garnish
- 1 bottle of ice-cold
  Schneider Weisse Meine
  Hopfenweisse

You can make this much simpler and just have Hopfenweisse and gin – honestly, it's great – but the addition of ginger and lemongrass syrup is worthwhile.

Put some ice in the glasses to chill them. Shake more ice and the syrup together for 30 seconds in a cocktail shaker. Measure the gin into the shaker.

Squeeze the lemon in through your hand to catch the pips. Add the beer and stir gently.

Empty the ice from the glasses and share the contents of the shaker between the glasses. Garnish with a twist of lemon peel.

*To make the simple syrup, add equal parts sugar and water to a saucepan. Bring to a boil and immediately turn off. Bruise the lemongrass and ginger and add to the saucepan, stirring to dissolve the sugar. Leave to cool and infuse. It will keep, in a sealed jar, for a few weeks.

3

Hop
Stars

Hops are like the perfumes of the beer world; these little plants bring with them more than 250 identifiable aromas as well as bittering compounds and anti-bacterial properties too, which means they are natural preservatives.

In fact, hops are pretty darn cool (and the closest known relative to cannabis, which is a good pub quiz fact). And it's in this chapter that we celebrate the best plant in the world, in my opinion. The plant that gives us aromas like rose, orange peel, grapefruit juice, sauvignon blanc, coconut, dill, lemongrass, key lime pie, cherry blossom, mango, lychee, pear and oh so much more.

In fact, you could say that this chapter is my love letter to Humulus lupulus and you wouldn't be far wrong.

Siren Craft Brew Lumina

ABV
4.2%

Country of origin
UK

Try it if you like
Fruity dry white wines

Great with
Grilled halloumi

Also try
Elusive Brewing
Oregon Trail IPA
5.8%, UK

Siren is based not far outside London, in what is a growing enclave of craft breweries: Double-Barrelled, Phantom and the truly excellent Elusive Brewing, run by the nicest-man-in-beer Andy Parker.

Siren has been going for just over eight years and has grown in strength to become a major player in the craft scene, with owner Darron Anley firmly at the helm and his love of big-flavoured beers writ large in the DNA of the brewery.

The beers have had a high quality about them since day one. There aren't many breweries that can boast this, but, in the last few years, they have become consistently excellent, and their barrel-ageing project is really beginning to hit its stride, with yearly small batch releases that are really worth looking out for. However, I still haven't forgiven them for discontinuing my favourite of their beers, Liquid Mistress, however.

If you're a coffee fan, then Siren is definitely a brewery for you. It has been running 'Project Barista' since 2013, to combine coffee and beer in as many inventive ways as they can come up with, often spending more on coffee in a beer than most breweries will spend on hops in an IPA, but the results are – even to a non-coffee drinker – truly stellar, although you have to be a bit careful about drinking too many as they come with a healthy caffeine kick!

All this aside, it's Lumina that I think is the real star of the portfolio these days. The plethora supple hops that weave around a deliciously light and refreshing body, has the perfect balance of bitter tang at the end and a soft sigh of tropical fruits that makes you think of a tangy sorbet and summer days.

Dougall's 942

ABV
4.2%

Country of origin
Spain

Try it if you like
Verdejo

Great with
Padron peppers

Also try
Hornbeer Top Hop
4.7%, Denmark

Brewery co-owner Andrew Dougall is a British expat who is properly living the dream. Opening up his brewery not far from Santander, he's as happy as a clam, or should I say happy as an *almeja*?

I firmly believe that he was dreaming of lying on the beach, as opposed to working in a hot brewery, when he came up with this beer because I can't imagine a better place to be drinking it.

A really simple two-malt, two-hop recipe, it's like a mix between a pale ale and a lager, making it perfect for its home environment or for when you're just dreaming of sun, sea, sand and session IPA (which, in my case, is about seven times a day).

Wild Card Brewery IPA

ABV
5.5%

Country of origin
UK

Try it if you like
Balance

Great with
BBQ brisket

Also try
Boneface
The Unit NZ IPA
6%, New Zealand

Wild Card has made quite the impact in its short history, much of which is down to head brewer and all-round superstar Jaega Wise.

A polymath is the best way to describe Jaega, as she is a talented presenter, scientist, brewer and musician. In fact, she'd be really annoying if she wasn't genuinely lovely and didn't make such damn good beer.

Based in Walthamstow, in North London, the brewery has proven to be a bit of a nexus for breweries, coffee roasters and other food and drink businesses.

Wild Card started in 2012 in the living room of two founders of the business, Andrew Birkby and William Harris, and has blossomed into a business that turns over more than a million a year and with listings in major supermarkets.

The IPA is a beer of theirs I return to frequently, because it's balanced, as most of the beers they produce are. There's no desperate hops arms race here, just a genuine understanding of the enjoyment of drinking a bitter, pithy, punchy IPA that won't laminate your tongue. Put simply, it's an ace brew.

Hop Notch
Hello World!

ABV
4.7%

Country of origin
Sweden

Try it if you like
Tropical flavours

Great with
The sea air

Also try
7 Clans
Hop-Rooted IPA
6.5%, USA

Co-owner of Hop Notch Jessica Heidrich and I go back a long way. My roommate when we are both judging out in America, Jessica is not only a brilliant scientist and brewer, but an absolute hoot to be around – and puts up with my snoring too.

Hop Notch was founded with her significant other Magnus, after many years of Jessica working for other people, and I couldn't be happier that she's got her own business, brewing in an old cinema on a beautiful island just outside Stockholm, with a view of the Baltic Sea. The projector room now houses the malt mill – you can look through where the projector was previously mounted to admire the shiny brew kit. And in the warehouse, the velvet curtains and silver screen are still in situ.

Hello World! was the beer they launched with and it sums up Jessica's love of hops perfectly; full of punchy Mosaic and Citra hops, bringing oodles of grapefruit and tropical fruit notes, and a lovely dry fermentation that makes it a very sessionable session IPA indeed.

Neptune Mosaic

ABV
4.5%

Country of origin
UK

Try it if you like
Lip-smacking refreshment

Great with
BBQ prawns

Also try
Bow & Arrow
Scenic West
6.5%, USA

Wife and husband team Julie and Les O'Grady make up the much-loved Liverpool brewery, Neptune, and have very quickly become absolute fixtures on the UK beer scene.

Julie was well-known for founding 'Ladies That Beer', a beer appreciation society, and as an outspoken advocate of stamping out sexism in and around pubs and brewing.

Les is less keen to step into the limelight but has had little choice but to do so since the brewery became such a hit – he still grumbles about it though.

Neptune is named after the family business of selling marine and tropical fish, and nearly all the beer names relate to mythology around the sea.

However, this beer does what it says on the tin. It is bursting with papaya and passion fruit, all layered masterfully on a light and smooth body, and an accessible ABV completes the picture – it's a real favourite of mine.

Big Smoke Cold Spark

ABV
3.8%

Country of origin
UK

Try it if you like
Pints with pals

Great with
Mapo tofu

Also try
Nomad Brewing Co.
Budgy Smuggler IPA
5%, Australia

I have to include this beer or suffer the indignity of the dog house at home for a while. It's something I frequently find in the fridge after my husband has been to the bottle shop – he adores it, and, although I have more fondness for their other easy-sipping pale ale Electric Eye, sometimes you just have to give in to the better half (I'm going to pay for this joke by the way!).

Big Smoke started in the back of the Antelope pub in Surbiton, which the brewery still owns, not far from where we live in south-west London, and was performing small miracles on a truly ramshackle kit. Although the consistency could be a little variable, the beers always held great promise.

Roll forward to now and the company is producing a lot of fantastic hop-forward beers, some pretty tasty gin, and a couple of pretty impressive dark beers too. It doesn't have an enormous range and it doesn't try to step outside of what it's good at. Plus it has a firm adherence to putting its beers in cask where suited, and it warms my heart to see that from a modern company.

It has moved on to larger premises slightly further out of the city, in leafy Esher, but is also now the owner of a pretty healthy stable of pubs and, as

I write this, is opening a bar in Heathrow. So, if you happen to be passing through Terminal 2, you'll be able to get a great pint, which is always a relief from the sea of beer mediocrity that normally greets you before a flight.

Cold Spark is simply a Citra hop-laden easy-drinker, with all of the bright sparky citrus and pithy bitterness that you could want from a beer touting this hop as its one and only.

2nd Shift Brewing
Little Big Hop

ABV
4.9%

Country of origin
USA

Try it if you like
Historic battles

Great with
St Paul sandwich

Also try
Epic Joose
6.5%, New Zealand

The folks at 2nd Shift are good ones, caring about their community, constantly giving back, and co-owner Libby Cridler is my kind of woman, taking absolutely no nonsense from no one!

Fortunately, they also brew really fine beers. It is so named because because co-founder Steve Crider fell in love with brewing and made it his life's work after his second shift.

A session New England IPA, Little Big Hop is stupidly easy to drink with waves upon waves of fresh pine, lime, tangerine and even a hint of mint right on the back end of the palate. It may have a rather combative hop on the label, but it certain doesn't go down with a fight.

Pohjala
Kosmos

ABV
5.5%

Country of origin
Estonia

Try it if you like
Tropical fruit

Great with
Feta

Also try
Boundary Brewing APA
3.5%, UK

'Passionate home brewers and beer enthusiasts' has become a trope for comedy sketches, with bearded dudes nerding out about IPAs and hops. But, as with all satire, it is both right and a little unfair on those who took this route.

Tallinn in Estonia has become somewhat of a beer-lover's destination, thanks in no small part to Põhjala. It was founded in 2011 by, yes, you guessed it, four dudes who were beer enthusiasts and home brewers. Enn Parel, Peeter Keek, Gren Noormets and Tiit Paananen were then joined by Scot Chris Pilkington. The Forest Series that the brewery produces is something I strongly recommend people try, it gives a real connection to Estonian folk lore and countryside, that is rare and to be appreciated.

However, I don't wish to tease you with only rare beers that become available in small quantities, it's the Kosmos I'm here to tell you about. When they describe it as being an 'intergalactic IPA' they aren't far wrong. The first time I tried this beer, the aroma nearly had me seeing stars! It absolutely reeks of tropical fruit (in a good way!), with wave after wave of mango, lychee, pineapple and a little hint of strawberry for good measure. The full, almost creamy body, with a hint of balancing bitterness, makes it almost like drinking an indulgent sorbet.

The Magic
of Hops

One of my most hated phrases in beer is 'hoppy'. You might hear it a lot
or might have wondered whether you're using it correctly. You might also
hear someone say 'oh, that smells hoppy' or 'you get a hoppy finish' and it
drives me bananas, which you might think is a bit weird and pedantic, but
let me explain.

The word hoppy is used as a lazy catch-all and, like many phrases that do
this, it is almost perfectly designed to keep beer novices in the dark. Why?
Because it means everything and nothing at the same time.

I have challenged some of the world's best brewers to tell me exactly
what the word hoppy means when I catch them using it in a tasting, and
not one of them has yet to give me a satisfactory answer (OK, yes, this is
pedantic but I think it's important to be annoying about this and my other
most-hated phrase 'malty').

Hops are one of the most powerful weapons in the brewer's arsenal.
They bring protection and stability to beer, offer a few nice antioxidants
to the drinker while they're there, and also bring a balancing bitterness
to the brew that stops it being overwhelmingly sweet... and that's
before we even begin to get to the enormous array of flavours and
aromas they bring to the party, which is why it irritates me to see
such a reductive description.

So, I explained a little bit about the plants and their history earlier, but I
just want to delve a little deeper into why they are so amazing. Got a beer
to hand? Good, this will get a little bit sciencey!

The Important Parts of Hops

Firstly, as you'll see from the illustration, there are different parts to a hop that give you different but very important elements in your beer.

Let's start with the strig. This connects the cone to the plant and has some tannic* qualities.

Moving onto the bracts. They also contain a bit of tannin but, most importantly, they form the outer layer to the structure in which the taste and flavour parts of the plant are held – the bracteole, which are home to the lupulin glands, and these, in turn, are full of the bittering-compounds in waiting in alpha acids, beta acids and essential oils.

The reason why I say 'in waiting' about the alpha acids, is because, at this stage, they can't be dissolved in water. This is one of the reasons why beer is boiled; it's so that they can go through a process called 'isomerisation', which creates iso-alpha acids which impart bitterness – see, warned you it was about to get sciencey! Beta acids are a little more complex. Their influence on a beer is based on time and oxygen and their contribution to beer and the perception of bitterness, while now better known, is still being fully studied.

*Tannins are a class of astringent polyphenolic biomolecules that produce a drying, puckering sensation in your mouth; if you think about underripe fruit or a very strong cup of tea, you'll be in the right area.

And finally, the bit most people get the most excited about, the essential oils. This is where the flavour and aroma compounds live, and where it gets really interesting.

The main groups of flavour and aroma compounds you get from hops are:

Myrcene
lemongrass, bay, cannabis, mango and thyme

Linalool
mint, bergamot, birch and rosewood

Farsenene
ginger, chamomile, grapefruit, lime and nutmeg

Geraniol
rose and geranium

Humulene
coriander, basil, clove, black pepper, sage or balsam

Beta Pinene
you guessed it, pine!

Carophyllene
black pepper, basil, and oregano or cannabis strains

Because the flavour and the aroma of the beer is what most people are more interested in these days (the fact that pale ales and IPAs continue to top the charts of best-selling craft beer styles is testament to this), science has moved on apace from when I wrote my first book *Let Me Tell You About Beer*, meaning hop extracts are now no longer only about getting bitterness and stabilisation in the beer. They have a really important role to play in modern brewing, and modern beer-drinking tastes, and are no longer dedicated to trying to stabilise a beer in the wrong colour glass bottle! (Top tip: don't drink beer that is designed to be 'drunk fresh' from a green or clear bottle; they will almost certainly have bad aromas and flavours.)

From CO_2 extractions of these delicate aromas and flavours (which makes them longer lasting and more stable in the beer) to the invention of Cryo Hops® (flash frozen at very low temperatures and then pelletised to seal in freshness), the world of hop processing has changed so much I could write a whole other book about it, but someone else far more qualified has already done that.

So, if you want to go full geek, get *For the Love of Hops* by Stan Hieronymus published by the Brewers Association – and, if you're doing that, you may as well pick up the malt, yeast and water ones too – because clearly, like me, you are now so deep down the beer rabbit hole, you are completely beyond help!

Brasserie du Grand Paris Nice to Meet You!

ABV
8.5%

Country of origin
France

Try it if you like
Lillet liqor

Great with
Smoked cod's roe

Also try
Kompaan
Handlanger DIPA
8.2%, Netherlands

Founded in 2011 by Fabrice Le Goff and Anthony Baraff, after a few homes – both brewing out of a tiny office unit and cuckoo brewing at Brasserie de la Vallée de Chevreuse – Brasserie du Grand Paris finally settled in Saint-Denis and has been flourishing ever since.

The area is probably best known for being the home of the Stade de France, which breathed new life back into what has historically been an area of some importance. During the Middle Ages, it was a centre of commercial importance, with a market that drew traders and merchants from all over the Byzantine Empire – and one can only imagine that quite a few deals were done over a mug of ale.

The brewery's retro artwork for all its beers is also extremely appealing, and they fly in the face of the very French tradition of labelling a beer by its colour – something I can really understand marks them out as something different. Instead of just 'blonde' or 'brune', they ensure that the consumers are fully informed on what they are buying with clear tasting notes – and Baraff's American heritage showing through on some of them, referencing things like graham cracker, which I'm sure causes some Gallic head scratching at times.

But, whatever the cross-cultural references, there is one thing that definitely shows through in all the beers I've tried from them, and that's how clean they are.

Nice to Meet You! in particular is an extremely well-balanced double IPA, packed to the gunwales with hops, driving through your palate like the tank on the artwork with oodles of tropical fruit and a proper pithy punch at the end. It elicits a 'très bien' from even the rustiest of French speakers like me.

I always think of it as one of those beers that you'd like to be sipping slowly, sitting outside a pavement café with a big pair of Audrey Hepburn-style sunglasses on, as condensation slips slowly down your glass and the hustle and bustle of Paris passes you by.

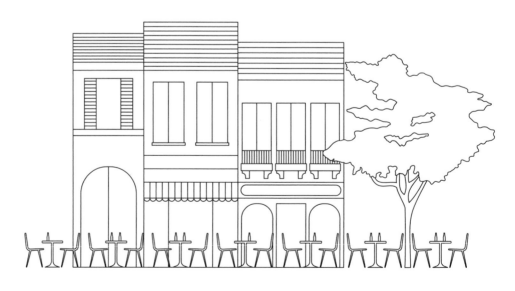

Two Birds
Bantam IPA

ABV
4.7%

Country of origin
Australia

Try it if you like
Long Island iced tea

Great with
Steak

Also try
Duvel Tripel Hop
9.5%, Belgium

I've been following the story of these plucky ladies from early on and I was thrilled when they finally got their bricks-and-mortar brewery in 2014. Since then have seen them go from strength to strength.

The two women at the helm, Jayne Lewis and Danielle Allen, have actually known each other since childhood, and it was when they embarked on two-week road trip along the US West Coast and their idea for a beer business emerged.

Two Birds is an obvious play on the Aussie (and British) nickname for a woman, but they have made it more than that. With their nest firmly established, they are kicking arse and taking names – as they should. Bantam is my personal favourite of theirs, a full-bodied IPA that shows restraint in both its alcohol content and its bitterness, while not stinting on its tropical aromas and pithy finish. In fact, this beer, like its namesake, punches well above its weight.

Brasserie de la Senne Jambe de Bois

ABV
8%

Country of origin
Belgium

Try it if you like
Vermouth

Great with
Charcuterie

Also try
La Trappe
Tripel
8%, Netherlands

I could write about every single one of this brewery's beers with love and affection every day of my career. Seriously, there is not a brew that they have made, that I've tried, that I haven't fallen instantly in love with. It's no wonder that co-owner Yvan de Baets is not only respected the world over, but he's also very well-liked too.

De Baets and his equally charming business partner Bernard Leboucq met in May 2002. They founded their first business towards the end of 2003, building a microbrewery in Sint-Pieters-Leeuw, in the former warehouse of the Moriau brewery, and christening it 'Sint-Pieter Brouwerij', which seems funny to me now, as it's such a traditional Belgian name for two such decidedly non-traditional men.

After producing there for two years, the premises became too small due to the pair's success. They then started working towards building a brewery in the city and changed name, to reflect their reverence for their home, to Brasserie de la Senne.

The move took a little longer than perhaps the pair would have liked, but they kept up production by renting spare capacity in other breweries and making sure they were hands-on the whole time.

They finally moved into their first premises in 2010, which is when I first met them as I was shown around the brewery as it was being built – I especially loved the 'cathedral to fermentation' room with its stained glass.

Fast forward to today and BDLS, as it is sometimes shortened to, is now in a state-of-the-art building and has moved towards being entirely organic, ensuring as little is wasted in the brewing process as possible and being as environmentally-minded as brewing allows.

The DNA that still runs through the brewery though is a love of balanced bitterness in their beers; Jambe de Bois being the greatest example. It would cause ructions at any judging table for the *tripel* style – to many people's minds too bitter to fall within the judging perameters – but that's what makes it great... and dangerous! Full of that honeysuckle floral you want from a tripel, it finishes dry as a bone, with a pleasing throat-catch of bitterness that makes you want to go back for more immediately, greedily even. But, do beware – it is strong and it will, like its piratical name sake, take you off at the knees!

Young Henry's Newtowner

ABV
4.8%

Country of origin
Australia

Try it if you like
Dancing at gigs

Great with
Live music

Also try
Philter XPA
4.2%, Australia

Young Henry's is an Australian stalwart brewery that is part of the fabric of the Sydney scene and is co-owned by one of my favourite hooligans in the industry, Richard Adamson.

The brewery is also responsible for engaging in what I think could be one of the most exciting ecological steps forward for the brewing industry that I've seen in my career, using algae to capture CO_2 from the brewing process and turn it back into oxygen. Working in conjunction with scientists at University of Technology Sydney Climate Change Cluster (C3), the brewery aims to lead the way for the industry in making brewing a more carbon neutral process.

Just to give you an idea, the CO_2 from the fermentation of just one six pack of beer takes a tree two full days to absorb and the bioreactor installed in the brewery produces as much oxygen as one hectare of Australian forest.

Newtowner is one of those beers that you want to bounce around at a sweaty gig with, as thirst-quenching as a can of lager, but with just enough pop of tropical hops and supporting sweet malt body to keep things interesting. It's sessionability at its finest.

Simple NEIPA Mango Sorbet

Makes 1 litre
(34 fl oz/4 cups)

- 2 large, very ripe mangoes,
  skinned and destoned
- 120 ml (4 fl oz/½ cup) water
- 125 g (4 oz/⅔ cup) caster
  (superfine) sugar
  (preferably unrefined)
- juice of 1 lime
- pinch of salt
- 120 ml (4 fl oz/½ cup)
  New England IPA

The fashion for New England IPAs, or NEIPAs, may be a very divisive one but there's no doubt the style is here to stay and, when done well, they can be very tasty indeed. With low bitterness and high fruit flavours, they lend themselves very well to this sort of sorbet.

Pop all the ingredients except the beer into a blender and blitz. Stir in the beer and, if you want super-smooth results, pass through a fine sieve or, if you're ok with a few lumps, don't bother.

Spoon into an ice-cream maker and churn until you have a sorbet. Alternatively, spoon into a tub and freeze, taking it out every half hour or so to rake the sorbet through with a fork until it's frozen.

NOTE: New England IPAs are the sexy new kids on the block. Hazy, with negligible bitterness and bursting with huge hop character, they are currently dividing the industry while bringing more drinkers into the fold.

The problem is, the very nature of them needing so many hops means that very few breweries actually have the means or the skill to make the same beer month in, month out, so I've chosen not to give you brands here, just advice.

You are looking for super-fruity smelling ones, you want to be overwhelmed with waves of mango, lychee, pineapple, apricot and all sorts of other stone fruits.

It's ok to have a hint of turps/diesel/pine but any more than a hint and you want to shy away from them for this recipe, as it won't work. If it's your thing though, go ahead and enjoy it – but it's not mine.

IPA and Grapefruit Vinaigrette

- 50 ml (1¾ fl oz/¼ cup) honey mustard
- 50 ml (1¾ fl oz/¼ cup) grapefruit juice
- 50 ml (1¾ fl oz/¼ cup) citrus IPA
- 1 teaspoon caster (superfine) sugar
- 250 ml (8½ fl oz/1 cup) groundnut (peanut) or grapeseed oil (or very light olive oil, but not extra virgin, it's too bitter)
- 1 egg yolk, per 500 ml (17 fl oz/2 cups) dressing
- sea salt and freshly ground black pepper, to taste

This is a really simple dressing that works very well with a variety of different salads, but my personal favourite is to put it with a warm grain-based salad like tabbouleh with some lightly poached fish or salty halloumi.

Put all the ingredients into an old, clean jar, tighten the lid very well* and shake until emulsified.

*I'm always prescriptive about putting the lid very tightly on salad dressing before shaking it; when I was a kid we had what is referred to as 'The Thousand Island Dressing Incident' where my dad forgot to check the lid on a bottle of said dressing and we were still finding blobs of it on the top of kitchen cupboards and other random areas of the room three years later.

4

Red, Amber and Brown

Poor brown beer, it gets such a bad rep. Nasty people call it 'twiggy beer' or 'old man's beer' or 'warm, flat beer' – they just don't appreciate its subtleties, so I've decided it needs a little bit of love before we start. So, here we go: 'I'm Melissa, and I love brown beer.' There, I said it!

I love a satisfying English bitter, with a bready, nutty body, a lightly hay-like, astringent middle and that balanced, lightly bitter but oh-so-moreish finish – I can honestly say that one of my most transcendent beer experiences was trying Bathams Best Bitter from cask for the first time.

But this chapter isn't all about bitters. I also enjoy a big, berry- and citrus-laden imperial red ale, stuffed to the gunwales with booming American hops and a dropkick of bitterness at the end, like one of my favourites: Bear Republic's Red Rocket. But sometimes I want a beer that's somewhere in between, maybe an amber ale with that refreshing quality of a bitter but with some of that brash hop character that IPA lovers lust after – like Wiper and True's Amber Ale. Hopefully this intro has inspired you to cast your net wider across this often neglected spectrum of beers (and yes, I have snuck three more recommendations in here because I really like these styles and I ran out of room for some of my favourites!).

Anspach & Hobday Ordinary Bitter

ABV
4%

Country of origin
UK

Try it if you like
Drinking in a proper pub

Great with
Another one

Also try
Bateman's
XXXB
4.8%, UK

It's hard to equate the two very nervous young men who came to my first book launch nearly 10 years ago to the burgeoning beer empire that is Anspach & Hobday today. Paul Anspach and Jack Hobday (yes, I know, they sound like a pair of Victorian detectives who foil the dastardly deeds of ruffians and ne'er-do-wells everywhere) have worked non-stop to build their business based on a some surprisingly traditional, as well as some excellently innovative, beers.

At a time when the beer world was obsessed with hops and in an arms race to make the most ridiculously strong and bitter IPA possible, they launched with their flagship London Porter, which I tried for the first time at the aforementioned event and was immediately blown away by, and they've barely tinkered with the recipe since.

Interestingly, as the beer market seems to be swinging back towards more traditional styles in the UK, and breweries globally are starting to re-embrace sessionable beers, they find themselves in the right place at the right time again with their exceptional Ordinary Bitter.

The name always makes me laugh, as it was a trigger phrase to one of the biggest characters in the UK scene when I first joined it, the late John Young. John was the head of the then London-based brewing empire Young's and was as eccentric, curmudgeonly, charming and brilliant as you'd want him to be – sadly he died not long after his beloved company was broken up and the brewery arm sold off.

However, John would take exceptional umbrage at anyone calling the company's pale ale (as it is still known internally by a lot of the old school publicans in the company) 'ordinary'. He would bark in his plummy tones 'My beer is ANYTHING but ordinary, thank you!' You'd only make the mistake once, that's for sure!

But, back to Anspach & Hobday's delightful beer. If anything, it bears more resemblance to a Fuller's beer than a Young's drop. With that rich, bready malt background and the incredibly restrained use of American hops. Instead of grapefruit you get marigold, and instead of biting spice you get warming black pepper. It is one of those 'down almost half the glass in one go' appeals that you want from a sessionable bitter.

Honestly, it's like the Pringles® of the beer world; once you pop that can, you can't stop.

The Wall Brewery
Sjavár Bjór

ABV
5.2%

Country of origin
Italy

Try it if you like
Salted caramel

Great with
Pork and 'nduja meatballs

Also try
Little Creatures
Rogers
3.8%, Australia

The Italian beer scene is, for me, improving so much from when I first encountered it. Holding so much promise for so long, with a handful of producers making exceptional beer, it failed to reach the heady heights of even a hundred excellent breweries nationwide until just recently.

The Wall initially collaborated on this beer with Birrificio Argo, a small brewery that also makes a very good American amber called Amberground, but the production has moved permanently to The Wall, as it's the larger of the two.

Based on the idea of, rather than a literal translation of, salted caramel, it's one of those beers that grows on you. The addition of sea salt right before packaging elevates the caramel notes in the malt body of the beer, makes the subtle hops a little brighter, and overall has the effect of making you reach for your glass a little quicker than you normally would!

Rascal's Big Hop Red

ABV
5%

Country of origin
Ireland

Try it if you like
Light Italian reds

Great with
Ligurian olives

Also try
Yellow Belly
Red Noir
4.5%, Ireland

Rascal is, quite simply, one of the finest words the English language, and this brewery produces some of the finest beers from the shores of Ireland, so it all marries up quite nicely.

Meaning to be naughty in a fun way, it does also seem to encompass how the brewery goes about its business, with a lot of fun social media and brand messaging and a cracking – almost dive bar – atmosphere at its pizzeria and brewery tap.

The beers are great, they really are, but they don't take themselves too seriously. Perhaps something that was picked up on co-owners Emma Devlin and Cathal O'Donoghue's travels in New Zealand, which saw them pack in their careers to come back home to Inchicore in Dublin and found the business.

There isn't a whole bunch of beer geek language used, there isn't a lot of confounding jargon, the beers are just spoken about simply and with affection. It was a tough call when choosing a beer from these folks, because the Yankee White IPA is also a personal favourite, but the big berry and piney, almost eucalyptus, notes in this beer just shaded it for me.

Bagby Beer
Three Beagles Brown

ABV
5.6%

Country of origin
USA

Try it if you like
English milds

Great with
A classic hot dog

Also try
AleSmith
Nut Brown Ale
5%, USA

Co-owner and brewer Jeff Bagby has an intrinsic understanding of how to make UK-style beers like few other US brewers I've met, balancing satisfying flavours with subtle complexity, and their Oceanbeach home is very high on my post-pandemic travel list.

Jeff and his business partner and wife Dande named Three Beagles after their love of the breed and it's a lovely, dry, easy-drinking brown ale with hints of toffee, dark chocolate and berry fruit, and a small suggestion of orange liqueur.

But be careful: it's one of those beers that only lets you know how strong it is about halfway down the third pint!

Fuller's 1845

ABV
6.3%

Country of origin
UK

Try it if you like
Malt loaf

Great with
Wensleydale cheese

Also try
Cambridge Brewing Co.
Arquebus
14%, USA

Fuller's may have been taken over by brewing giant Asahi now, but 1845 remains a masterpiece of the bottle-conditioned art.

Commissioned to celebrate the brewery's 150th birthday, it was so popular it stayed and each bottle is aged for 100 days before being released from the brewery – something the modern purveyors of dangerous exploding fruit beers in cans could perhaps learn from.

Bottle conditioning means it goes through a secondary fermentation in the bottle, which adds not a lot more alcohol, but definitely a lot more complexity. It also has the added benefit of scavenging the oxygen collected in packaging, making it a lot more shelf stable.

Put simply, this is a fruitcake in a glass, and it really does deserve to be poured into a bowl shaped glass of some sort in order for you to completely enjoy its aroma.

Masses of raisins, prunes and apricots are joined by the soft breadiness of the amber malt and a light dusting of fresh hay from the traditional Goldings hops.

More About Malt and Why It Matters

It's so refreshing to see beers with proper malt flavours come back into vogue. It seems craft beer is exiting its years of brewberty, after everyone has had their flirtation with loud, obnoxious but somehow attractive triple IPAs, had a few weird nights with some face-changing sours and comfort-drunk five blueberry jam cronut chocolate stouts in a row like they've just had a bad break-up and are now looking for dependable beers that are balanced and soothing. Beer you want to drink for life.

Tortured comparisons to teenage and adult life aside, the story of malt and how it came to be so integral to beer-making literally starts before the dawn of civilisation... in fact, without grains there might not even be civilisation.

Ever-increasing amounts of archaeological evidence shows that where there were pockets of grain-growing country, people would settle down from their nomadic lifestyle to tend to the crops. From there, folks had to start working out a way to live together and not whack each other over the head with a club every time there was a small disagreement.

In fact, it's highly likely that we are predisposed, or in fact genetically coded, to like alcohol. When we were still swinging about in trees, fallen fermenting fruit would have been a powerful enticement to head to the ground. Rotting (i.e. fermenting) fruit has a much stronger aroma, is easier to digest and is a readily available source of calories, giving you the jump on the competition in terms of food and energy.

In a fascinating article in *National Geographic* in February 2017, titled 'Our 9,000-Year Love Affair With Booze', one of the most telling parts about how alcohol enjoyment is hardwired into us is that there was a critical gene mutation in the last common ancestor of African apes and us, which geneticists recently dated to at least 10 million years ago. The mutation allowed that ancestor to produce an enzyme to process ethanol 40 times faster than we could previously achieve, showing how important this source of energy was.

There is such a rich history of alcohol and its role in society. From the earliest discovered alcohol in Jiahu, China, where by 7,000 BC farmers were already fermenting a mix of rice, grapes, hawthorn berries and honey in clay jars, to today's all-singing, all-dancing brew kits. It would be easy to spend pages and pages rhapsodising about its role, and discussing the ways that society has attempted to deal with its excesses, but I do need to get back to malt itself.

Malt, as mentioned earlier in the book, is vital to the brewing process itself. It's a wondrous bundle of booze potential, containing an enzyme that will turn starches into sugars that, in turn, will be turned into alcohol by the yeast.

The best growing conditions for malting barley are maritime climates, of which the UK is one, making British malt some of the finest – if not the best – in the world.

The processes for kilning and drying the grains can vary. There are some traditional maltsters who use floor maltings still – careful heating and drying the grain and raking it over by hand – to produce a truly artisan product that some believe to be superior due to its gentle treatment. Other maltsters use huge drums for this process, kind of like industrial tumble dryers if you will, which gives them incredibly precise control over the levels of humidity and heat applied to the malt, resulting in a very consistent product. (I am not weighing in on either side by the way, I like my life too much!)

The two most-often used grains in brewing are barley and wheat, but you might also see oats, rye, spelt, sorghum, corn/maize and other grains.

The main point of malting is about enzymic potential: the point where the moisture, pH and temperature hit the sweet spot in the malted grains that make up the majority of any beer's base. This is when the maltster halts the process, allowing the barley to release all that sweet, sweet sugar. It also acts on other grains too, releasing any of their fermenting potential.

And finally, what that allows the brewer to do is figure out how much alcohol they will end up with when they've let the yeast do its thing.

According to one translation of a play by the ancient Athenian comic poet Eubulus, Dionysus, the god of drinking, issues this warning:

I mix three kraters only for those who are wise.
One is for good health, which they drink first.
The second is for love and pleasure.
The third is for sleep, and when they have drunk it those who are wise wander homewards.
The fourth is no longer ours, but belongs to hubris.
The fifth leads to shouting.
The sixth to a drunken revel.
The seventh to black eyes.
The eighth to a summons.
The ninth to bile.
The tenth to madness,
in that it makes people throw things.

North End Brewing Amber

ABV
4.4%

Country of origin
New Zealand

Try it if you like
Marmalade

Great with
Spicy pork ribs

Also try
Jopen Jacobus RPA
5.5%, Netherlands

Brewer Kieran Haslett-Moore is a man with a serious passion for beer. Formerly a beer buyer for an off-licence chain, he has successfully done what few do: switched sides, and he has been gaining fans ever since.

I had the pleasure of a collaborative brew day with Kieran. Amber was a beer that was drunk a fair bit while we used foraged ingredients from his late grandfather's lands, shot the breeze about our ethos around beer and generally had a great day.

The reason the Amber was a beer we returned to was its easy-drinking nature, which just allowed conversation to flow (and was also particularly satisfying after digging out the mash tun!).

Taking influences from both US and UK bitter styles, it is still a uniquely New Zealand beer due to the hops used. It contains beautifully integrated caramelised marmalade and rye bread hints, with a suggestion of thyme and lime zest, and is deeply refreshing.

Texels Bock

ABV
7%

Country of origin
Netherlands

Try it if you like
Oloroso sherry

Great with
Smoked ham hock
and mash

Also try
Ayinger Celebrator
Doppelbock
6.7%, Germany

On an island off the north of the Netherlands is a population that is outnumbered by sheep about three to one but, amazingly, its brewery is about to open a second site dedicated to brewing its flagship beer - that's a success story if I ever heard one.

And Texels brewery is all about passion, from its man-mountain ambassador Hans Glandorf, who lives, breathes and bellows the brand, to how they source their ingredients, to supporting the local economy.

Although Bock is technically a lager, it has more in common with a strong ale on the palate with its richness. The first thing you'll notice is the striking deep amber colour with garnet highlights, so it's worth pouring in a big red wine glass to truly admire it. Then the nose hits you, big boozy, raisiny, pruney Olorosso sherry notes interwoven with some orange peel and a brisk carbonation that make it both complex and dangerously drinkable at the same time.

It's also worth noting that the Dutch Bock Bier festival in Amsterdam is the largest beer fest in the whole of Europe dedicated to just one style... got to love those crazy Dutch!

Sierra Nevada Torpedo Extra IPA

ABV
7.2%

Country of origin
USA

Try it if you like
Sierra Nevada Pale Ale

Great with
Slow-cooked lamb shoulder

Also try
Pirate Life IPA
6.8%, Australia

No beer book would be complete without a mention of the truly pioneering and magnificent Sierra Nevada.

The brewery, or should I say breweries, as it now has two sites, is still family-owned and run and is estimated to be worth a whopping billion dollars.

But mere money can't put a price on its contribution to the resurrection of great beer the world over. From in-house childcare to a commitment to environmental consciousness, it is a business model to which all others should aspire.

Torpedo is a mainstay of the brewery's portfolio and, despite its weighty 7.2% ABV, is dangerously drinkable. Filled with pineappley, grapefruit notes and a chewy caramel body, it could easily sink you if you don't respect it, so do be careful!

Uerige
Altbier Classic

ABV
4.7%

Country of origin
Germany

Try it if you like
Pretzels

Great with
Pretzel and cheese dip

Also try
Duckstein Altbier
3.7%, Australia

I am a huge fan of Altbier; in fact, I love it so much that I keep pestering a local-to-me brewery to bring it back to their list. Sadly, I appear to be a fairly lonely lover of this style in the UK's capital. Luckily though, I can take solace in this truly excellent version.

It's extremely rare that you'll find a bottle-conditioned beer in Germany. This is a process where you make the beer in the usual way, but then you give it a secondary ferment in the bottle with either the residual yeast in the beer or by adding a little dose to the bottle with some fresh yeast. This does a number of things: firstly, it very efficiently scavenges any excess oxygen in the bottle, which is good for keeping the beer fresh, and, secondly, it also creates a smoother carbonation with some extra flavour as a bonus.

This classic German variety is probably the most robust and full-flavoured of any of the traditional Altbiers still being produced in Germany. It has a lightly gingery note, huge, soft pretzel middle and a peppery spice at the end that makes you cluck your tongue in delight.

Anchor Steam Beer

ABV
4.9%

Country of origin
USA

Try it if you like
Fresh bread with marmalade

Great with
Mission-style burrito

Also try
Hammerton
Islington Steam Lager
4.7%, UK

There is so much to say about Anchor Brewing in San Francisco – not least about its owner from 1965–2010, Fritz Maytag, and legacy as a forefather of craft beer in America.

You may recognise the Maytag name, as the family were inventors of the modern-day washing machine. A 25-year-old Fritz, seeing that the brewery's fortunes were on the wane, decided that he would stake much of his inherited wealth on reviving its fortunes. Flying in the face of the 60s stock market 'bear' state, he gave away more and more of his stock to save the brewery, much to his family's dismay. Eventually, however, his investment in positioning the brand as a premium product paid off, and it went on to become an American icon.

There are many legends and myths around why it's called 'steam beer', which is also now known as California Common, but what can't be argued are its roots in the German Altbier style. The rich, malt body is instantly recognisable as such, but there's a little extra tang from the Northern Brewer hops grown in the US, which add an almost over-blown orange peel punch with a hint of nettle.

Bacchus
Oud Bruin

ABV
4.5%

Country of origin
Belgium

Try it if you like
Amontillado sherry

Great with
Smoked almonds

Also try
Liefman's
Goudenband
8%, Belgium

Kasteel Brouwerij Vanhonsebrouck has spent most of its history bucking trends. From strong women at the helm, to fighting Brussels for the right to call some of its beers *gueuze* – one of only two businesses outside the Belgian capital allowed to do so by the EU – it's quirky, family-owned and housed in a deeply impressive building.

With a history that dates back to 1865, to a farm brewery in the village of Werken, there have been a few moves and family power shifts along the way. The brewery really took off after WWI, when the aptly-named Louise De Poorter, who had been running the brewery as well as taking care of five children (the history conveniently glosses over why her husband Emile wasn't doing so), leaves it to her ambitious son, Paul. Paul, along with his sibling Ernest, started making aggressive expansion plans, and their own maltings – with the builders paid half in money and half in beer – until disaster almost struck the family dynasty with Paul becoming ill and dying relatively early and Ernest and his wife not having children.

However, one of Paul's older sons, Luc, decided to take up the family mantle. Despite a previous switch into the production of Pils, in 1956 Luc decided to roll back the chasing of modern beer fads, realising he couldn't compete with the big boys, and started to focus on the old style of brown ale, calling it Bacchus.

It was a bold move, and one that led to a gueuze war with Brussels producer Belle-Vue (now owned by AB InBev) – even to the point where they sponsored rival football teams' jerseys. The brewery fought against the Brussels-led decree that they couldn't call one of their beers, St Louis, a *gueuze* – and won.

The Vanhonsebrouck family is still the proud owner of the Ingelmunster Castle, which has not only inspired the hugely successful Kasteel range of beers, but also the rather imposing fortress shape of its enormous brewery.

The Bacchus Oud Bruin is often derided as a 'training wheels' *oud bruin*, but it's for that very reason that I wanted to include it. I think it's a wonderful introduction to what can be a truly rewarding style of beer. It's not one to smash back in my opinion (although, you do you), it's a beer for spirited conversations or quiet contemplation, with its interplay of deep red fruit, slight leathery dryness and a hint of chocolate and caramel. It's also a truly delicious accompaniment to some aged Gouda.

Ska Brewing Pinstripe Red Ale

ABV
5.2%

Country of origin
USA

Try it if you like
Blackcurrants

Great with
Croque-madame

Also try
Good Chemistry ESB
4.3%, UK

There is something in the air in Colorado that makes so many damn good brewers – seriously, I don't know if it's the Rockies or that so many of them are nearly 7ft tall, but there's something going on in that state that seems to lend itself to great beer and Ska is no exception.

Run by joyfully anarchic characters, it is nonetheless very serious about its beer – although in a rather irreverent way. While I disagree with the brewery's characterisation of a 5.2% beer as 'sessionable', I don't disagree that it's brilliant, so that's why it's here!

A soothing caramel base is very simply overlaid with Liberty hops, which add a tinkly blackcurrant, some rich lily-like notes and a nice, nettly spice that brings everything to a sweeping crescendo at the finish.

Black Bean Dip with US-style Pale Ale

Serves 6

- regular olive oil
- 3 medium-sized white,
  brown or yellow onions
  (300 g/10½ oz), roughly
  chopped
- 4 garlic cloves, crushed
- ½ teaspoon ground cumin
- 1 tablespoon Mexican
  oregano (or regular if
  you can't get Mexican)
- ¼ teaspoon asafoetida
  (hing) (optional)
- 6-10 slices of jarred,
  pickled jalapeños,
  roughly chopped (depends
  how hot you like it)
- 2 × 400 g (14 oz) cartons
  or tins of black beans,
  drained but not rinsed
- ¼ teaspoon fine sea salt,
  plus more as required
- 50 ml (1¾ oz) US-style
  pale ale
- grated zest and juice
  of 2 unwaxed limes
- small handful of
  coriander (cilantro)
  leaves, finely chopped,
  to garnish (optional)*

*Fresh coriander (cilantro) has
been shown to be a genetically
divisive flavour, hence the
optional aspect.

After only ever having had terrible meals of pastiche Tex-Mex food in London, it wasn't until my much-missed, late friend Glenn Payne introduced me to the delights of proper Mexican food that my eyes were opened. He even gave me a great cookbook on the subject as an unexpected present one day. I'd like to think he'd hoover this up with gusto.

Put enough oil to cover the base of a medium-sized saucepan on a low heat, add the onions and cook for 10 minutes until softened and translucent, stirring from time to time.

Add garlic and cook for a further 5 minutes, stirring occasionally.

Add the cumin, oregano, asafoetida, if using, and jalapeños and stir well. Add the beans. Cook gently for 5 minutes, stirring regularly.

Allow to cool for 10 minutes, add the salt and blend with the beer and lime zest (using either a stand or handheld blender) until smooth.

Add half the lime juice and stir, then taste, and adjust the seasoning and acidity accordingly.

Spoon the dip into your chosen serving bowl and garnish with coriander leaves, if using.

Simple
Beer Bread

Makes 1 loaf
(23 cm × 13 cm/9 in × 5 in)

For the dry ingredients

– 300 g (10½ oz/2 cups)
  strong wholemeal flour
– 200 g (7 oz/1⅔ cups)
  strong white flour
– 1 teaspoon fine sea salt
– 1 teaspoon dried yeast

For the wet ingredients

– 1½ teaspoons maple syrup
– 150 ml (5 fl oz/⅔ cup)
  boiling water
– 150 ml (5 fl oz/⅔ cup)
  sweet brown ale
  (or Dunkelweizen)
– 1½ teaspoons groundnut
  (peanut) oil (or another
  fairly neutral oil)

You can use a stand mixer or your hands for this. Either is fine – the timings are all the same. You'll need a ½ pint (285 ml/9 fl oz) of tap water to hand for the baking part.

Put all the dry ingredients in a bowl or mixer and mix together.

Add the maple syrup to the boiling water and stir until dissolved, then add the cold beer. This should give you the perfect temperature to make the dough.

Stir half the liquid into the dry ingredients, then continue to add smaller amounts until a dough forms. You may not need all of it, you may need a bit more – every flour is different! Knead for about 8–10 minutes until smooth. Turn out into a lightly oiled bowl, cover with a damp tea towel and leave somewhere at 26°C (80°F) for 30–35 minutes.

Preheat the oven to 210°C (410°F/Gas 8).

Knock back the dough and leave to rise again for another 15–20 minutes. Gently turn the dough out into a lightly oiled loaf tin and get ready to move quickly!

Put the bread tin within grabbing distance and pick up the glass of tap water (see intro above). Open the oven door and throw the water on the base of the

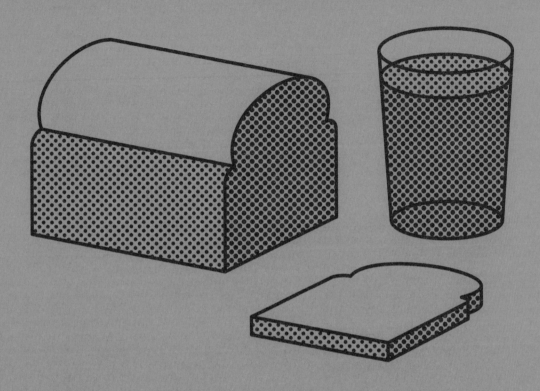

oven, grab the tin and put it on the middle shelf and close the door as quickly as possible – this helps form a good crust on your loaf.

Check the loaf for even browning after about 25 minutes and turn it if necessary. It should be done in about 40 minutes total.

Leave for at least 25 minutes. I know it's torture, but that just completes the final bit of cooking in the middle and means you won't get a tummy ache!

5

Farmhouse Beers

Farmhouse beers really refers to most beers that were brewed before the industrialisation of the process. Chucking fermentable ingredients in a pot and boiling them up to make booze is pretty much as old as upright mankind, but in the beer world it mainly refers to the Belgian and French styles of farmhouse brewing.

The main difference between the French style of farmhouse ale, known as Bière de Garde, and the Belgian, known as Saison, is the ABV and the different characteristics of the yeast strains.

Bières de Garde tend to be stronger and akin to a blonde ale, while Saisons tend to be drier and leaner and have a more peppery, spicy character, but they do have something in common in that they were working-class drinks. Made mainly for migrant farm hands, they were often the result of whatever leftover crops were fermentable, then left to age until the next harvest.

Saison Dupont

ABV
6.5%

Country of origin
Belgium

Try it if you like
Bone dry white wine

Great with
Smoked foods,
like tofu or duck

Also try
Wild Beer Co.
Ninkasi
9%, UK

Quite simply, Saison Dupont is the standard by which all other Saisons are judged. Historically, perhaps, this might not be a truly accurate representation of a Saison, or perhaps it is – it's hard to really know due to the lack of recordkeeping. But, what it is fair to say is that modern classic Saisons, like Kansas City's Boulevard's Tank 7, have the DNA of Dupont running through them.

Meaning 'season', the Saison style of beer has been mucked about with, had all sorts of fruits and spices and other ingredients added to it by multiple breweries to great or lesser success – I've even collaborated on a few myself – but there is something incredibly pleasing about returning to the original again and again.

The site of the brewery was a farm dating back to 1759, and as far back as the mid-1800s it was known for its beer and honey production.

The beers really came to the fore, however, in the 1920s, when, to dissuade his son Louis from moving to Canada to buy a farm, Alfred Dupont bought him the farm-brewery. Things progressed nicely until the first and second World Wars ravaged the continent of Europe and beyond.

Add to this the rise of Pilsner in post-war Belgium and the brewery skated close to extinction,but, under the stewardship of Sylva Rosier, the introduction of Redor Pils saved the day.

I could go on and on about the history of the brewery – it is fascinating. It remains in the hands of the Rosier family today, who have diversified with laboratories for ingredient analysis, a cheese-making business, a range of organic beers and even, introducing a bit of modernity, a dry-hopped version of their flagship Saison.

But it's the original I return to like an old friend time and time again. It is without doubt one of the most versatile weapons in my food-pairing arsenal, but it can be enjoyed as a simple sipping beer just as easily.

Peppery, dry, lightly fruity, so highly carbonated it makes your nose tingle and wiggle like a rabbit, it is, without doubt, a beer I fear I could not live without.

Le Baladin Wayan

ABV
5.8%

Country of origin
Italy

Try it if you like
Gewürztraminer

Great with
Steak with green
peppercorn sauce

Also try
Birra del Borgo
Duchessa
5.8%, Italy

The mad professor of Italian brewing (which is saying something!) Teo Musso, the founder of Baladin brewery, continues to pick up awards and turn out spectacular beers in equal measure.

Based in the Piedmont region, he is as entrenched in food as he is beer, and Wayan is just one of his beers that demonstrates that with 17 different ingredients, including five grains (spelt, buckwheat, rye, barley and wheat) and nine spices, of which five are peppers – and none of which dominate.

On the nose you get pear and Earl Grey tea, with a hint of white pepper and coriander (cilantro), then on the palate it's lightly oily, with a hint of Cointreau and a bright, spicy finish.

Burning Sky Saison À La Provision

ABV
6.5%

Country of origin
UK

Try it if you like
Fino sherry

Great with
Smoked cheese

Also try
Anything from Scratch Brewing, an ever-rotating foraged beer brewery in Illinois, USA

When Mark Tranter announced he was leaving Dark Star and starting his own brewery, the whole UK beer scene sat up and took notice. He walked me round the place where he was hoping to build, and it's one of the toughest things I've ever been sworn to secrecy on. I wanted to shout from the rooftops that this was going to be something special.

And special it is. While waiting for planning permission, Mark went on a whirlwind odyssey, visiting some of the best brewers of farmhouse ales, from Colorado to Belgium. Armed with tips, he came back and opened Burning Sky and neither he nor his fans have ever looked back. Although he doesn't focus solely on Saisons – he also makes some fantastic session beers and IPAs – you can see in his eyes that the foeders in his brewery are his biggest love.

Saison à la Provision is brewed with Mark's purist hat on. A traditionally hopped wort is first fermented with a Saison yeast and then treated to a dose of Brett and Lactobacillus for a dry, tart finish to the straw-like, bready body with a hint of marigold. If I were a farm worker given this at the end of a long hard day, I'd probably kiss my employer there and then!

Brasserie Duyck Jenlain Ambrée

ABV
7.5%

Country of origin
France

Try it if you like
British bitters

Great with
Tarte Tatin

Also try
Wild Beer Co
Wild IPA
5.2%, UK

As it's been brewed since the 1920s, I think we can definitely call Jenlain Ambrée a classic.

A proudly independent brewery, Brassiere Duyck has remained in family hands, transforming over time but never changing its ethos of creating local beers first and foremost. The all-important ageing stage of around a month at a very cold temperature gives Jenlain its clean flavour, disguising the surprising level of booze!

With its aroma of crunchy autumn leaves, steamed orange pudding, warm spice and an almost savoury celery salt note, it's a delight for all the senses and one that I don't think gets enough attention.

Wildflower Brewing and Blending Good as Gold

ABV
5%

Country of origin
Australia

Try it if you like
Pet Nat

Great with
Rind washed cheese

Also try
New Belgium
La Folie
7%, USA

From the off I have to warn you that not one of these beers is the same but, having had the privilege to try one of the early versions when I was in Australia, I can honestly say that I sincerely doubt you'll be disappointed in any of the blends that are released on a regular basis.

I first crossed paths with one of the founders, Topher Boehm, when he was working for Partizan Brewing in my home city of London, and it's no surprise to me that an alumni of that brewery settled into a quirky brewery in Australia, Partizan's attitude to not chasing the hype may perhaps have played a role.

He has an incredible pedigree, having worked at the renowned US company Jester King and his now neighbour Batch Brewing, where the brewery sources its wort for ageing and blending.

There is a deep sense of science and philosophy around the project, from spraying the new brewery with inoculated wort, so that the wild yeasts that create the complex fermentation characteristics would imbue the very air and surfaces around the barrels, to the solera systems installed and the blending done on a purely sensory level.

It's not a particularly financially rewarding way of brewing to be quite honest with you, and people who don't understand the sheer passion, dedication and immersive nature of these beers might sometimes be utterly baffled by the price tag that comes with this style of beer. They will never be for everyone, but I would honestly get on a plane and fly back to Sydney in a heartbeat to drink Wildflower beers... and it's something I hope to do again soon.

While it's difficult to put truly accurate tasting notes on the Good as Gold, as each batch varies, what I can tell you is that the fermentation characteristics that are teased out in the beers I tried were absolutely on the right side of funky; dry but not astringent, leathery but not goaty, and very fruity, with sun-warmed dried apricots being the abiding memory I have of it. And please, don't send me pictures on social media if you buy one – it will only make me sad I can't have any!

A Quick History of the Use of Plants and Other Things in Beer

Brewing's roots lie with women in the Middle East and Africa, although sadly much of this rich heritage has been lost to a lack of literacy, women's erasure from history and the slave trade.

As humans, we have literally developed on an evolutionary level to process alcohol better and, let's be truthful here, that's mostly because mankind has always tried to make alcohol – and other conciousness-altering substances – from anything that even faintly looked like it might do the trick.

There are so many things that have been used in beer over time. From dates in ancient Egypt, to the chewed up corn used in chica in Peru, to millet for t'ella in Ethiopia, to the plants used prior to the introduction of hops like mugwort, alecost and yarrow. It's a manufacturing process that owes much to both tradition and innovation, which both play a part in the incredible range of ingredients that are used in beer today.

Whether it's the aforementioned date beer in Egypt, cherry beer in Britain and Belgium, or pumpkin beer made by early American settlers – if it adds flavour and contains starches or sugars that will ferment and create alcohol, we've brewed with it.

On the African continent and in South America, for example, use of the cassava root as the basis of fermentation is also being more widely explored, as brewers are seeking a more sustainable and localised way of brewing beer that isn't solely reliant on imported barley.

And, of course, you can't forget the use of honey or rice for beer's close neighbours sake and mead or rule out the likelihood that a collection of many of the aforementioned ingredients weren't chucked into the mix when farmers were making beers as part payment for their seasonal labourers.

As we continually discover more about brewing, and human, history, modern brewers are resurrecting or reimagining what came before.

Duration
Bet The Farm

ABV
4.5%

Country of origin
UK

Try it if you like
Dry Riesling

Great with
Sitting on a hay bale

Also try
De Ranke
XX Bitter
6%, Belgium

Duration Brewing was one of the most hotly anticipated openings in the UK in the past few years. After a short period of contract brewing their beers elsewhere, the dynamic duo of dyed-in-the-wool, hyper Londoner Miranda Hudson and softly-spoken, deep southern American, Bates (honestly, there's no point in typing out his full name, not a soul in beer would know who you are talking about!), realised what, for many, is the dream brewery in the backwaters of Norfolk.

Based on the site of an old priory, the pair have created a destination farmhouse brewery with a 20 hectolitre brewhouse on the banks of the River Nar. Having overcome worries about bat colonies and archaeological hurdles (I'll never forget the text I got from Miranda during the initial searches with the non sequitur of 'no bones'), the brewery got up and running just before the global pandemic hit.

However, already having built interest from the beers that they were brewing elsewhere, the IPA Turtles All the Way Down being a firm crowd favourite, and their commitment to using both local ingredients (it does help that Norfolk grows the best malting barley in the world), global hop varieties, modern stainless steel and old farmhouse brewing techniques like foeders (big wooden barrels), they are fusing modern and traditional beers with their modern-traditional vision for their business and lives seamlessly.

In future, there will be two versions of this beer. One 'fresh' version, made in stainless steel, will offer a delicious expression of herbal, spicy Continental hops on a peppery pale ale. It will temporarily confuse your expectations – it certainly did mine. My exact original tasting notes were: 'It's a bit of a cloudy pale ale. Oh, it's fresh like a lager. Oh, where did those fruity notes come from? Wow, that's dry. Oooooh, herbal!'. I'm a professional, honest!

The second version will be foeder-aged and will have all the complexities of a mixed fermentation beer but with the same DNA running through it, so look out for its limited releases.

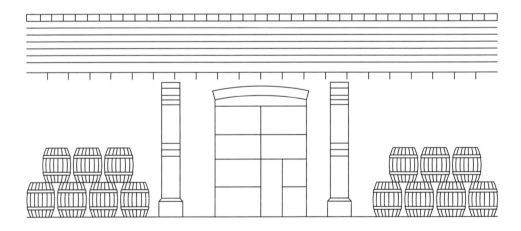

La Choulette
Bière des Sans Culottes

ABV
7%

Country of origin
France

Try it if you like
Natural sparkling wine

Great with
Goat's cheese

Also try
Le Baladin Wayan
5.8%, Italy

Although directly translated as the beer without trousers (pants), this beer's name refers to the fact that, during the French Revolution, the revolutionaries wore long trousers rather than the culottes, or knee-breeches, that were worn by the upper classes against whom they were rebelling.

In 1895, Jules Dhaussy set up a small brewery on his farm, using the barley he grew. Jules' son, Alphonse, took over, but in the 1950s he gave up brewing, partly owing to health problems and partly because his brewery needed major investment. He kept up the farm, and his eldest son, also called Alphonse, went to work for a brewery in Valenciennes. Fast forward to 1977: Alphonse found an artisanal brewery was up for sale just 1.9 miles (3 km) from his family's farm, bought it and appointed his son Alain to run it, having renamed it La Choulette.

It's definitely a beer with its big person pants on, tea-like earthy freshness and lovely honeyed sweetness. Then, when you take a well-chilled sip, you get a flavour like cloudy honey spread thickly on fresh bread. This is then joined by a pleasing zesty Earl Grey note with the citrus-sweet presence of lemon curd and an ephemeral dry pear note at the end.

It's Always Crumpet Saison

Makes 8-10

- 350 ml (12 fl oz/1⅓ cups) warm full-fat (whole) milk
- 1½ teaspoons golden caster (superfine) sugar
- 200 g (7 oz/1⅔ cups) strong white flour
- 150 g (5 oz/1¼ cup) plain (all-purpose) flour
- 4 teaspoons dried yeast
- 200 ml (7 fl oz/scant 1 cup) warm Saison such as Wayan
- 1½ teaspoons salt
- 1 teaspoon bicarbonate of soda (baking soda)
- flavourless oil, for cooking
- salted butter, to serve

Crumpets smothered in butter are one of the world's finest comfort foods, and I am genuinely chuffed to bits at the discovery that adding a little beer to the process makes them even more chewy and awesome. You will need metal cooking rings and a heavy-based frying pan for this – nothing else works as well.

Put the milk and sugar in a saucepan and start to warm over a very gentle heat.

Sift the flours and yeast into a bowl. Once the sugar has dissolved into the milk (which should be just hot enough that you can stand to put your pinkie into it, no hotter), pour it over the flour and mix briskly for about 5 minutes using a wooden spoon or stand mixer. Cover and leave to stand for 30 minutes– 1 hour in a warm place.

Once the batter has risen and started to fall again, pop the beer and salt in a pan and start to gently warm. DO NOT LET IT BOIL. When the salt has dissolved and the beer is about the same temperature as the milk was, add half of it to the batter and then sprinkle the bicarbonate of soda on top. Mix quickly and keep adding more of the warm beer until you get a nice runny batter about the consistency of single (pouring) cream. Put back in the warm place for at least 30 minutes. It should be pretty bubbly before you start cooking.

Heat a heavy-based pan over a very low heat for at least 10 minutes so it's evenly hot. Grease the inside of the metal cooking rings. then place in the pan to heat up.

Fill the rings with batter to about a third of the way up and leave, undisturbed, until the surface has set. After about 5 minutes you should have a gloriously bubbly surface. Carefully turn the rings over and ease a palette knife around the edges to release the crumpets. Cook on that side for a further 5–8 minutes or until just golden brown.

Repeat until you've used all the batter. Toast lightly, slather with butter and fill face!

Saison Julep

Serves 2
(in tumbler-style glasses)

- 300 ml (10 fl oz/1¼ cups) cold Saison such as Brew by Numbers Motueka & Lime Saison (if it's flavoured or highly hopped, make sure it's citrus)
- 2 × 50 ml (1¾ fl oz/¼ cup) shots of decent bourbon (I use Four Roses)
- 10 mint leaves
- 2 teaspoons golden caster (superfine) sugar
- crushed ice

A Saison really lends itself to a julep, making sure that it's brisk and refreshing, and adds a little more body than the usual soda water.

Divide the Saison between the glasses. Pour a shot of bourbon into each glass.

Muddle together the mint leaves and sugar and stir into the Saison and bourbon until the sugar is dissolved.

Spoon in the crushed ice, stir softly and top with a fresh mint leaf. Serve with some short straws.

6

Wild and Tamed Ones

The most traditional examples of 'wild' beers are considered to be the Lambics of the Brussels area of Belgium. Protected under law, only beers produced in that area* can be called Lambic – as a result, you'll see beers from other areas of the world call themselves 'Lambic-style'.

But, as with everything, clever people find a way to recreate this style all over the world, whether it's in Cambridge in the UK or San Diego in the USA – these traditional methods are being resurrected after being moth-balled for years.

All-in-all, it's a tricky category to navigate because it heads into the world of the weird cousins of regular beers – you know the kind, the ones that you only invite to one party a year and even then they try to eat the candles on the birthday cake. So, with that rather worrying image in your mind, let's head off into the realms of wild yeast and bacterial fermentations.

Before you go, you'll see I talk about something called 'coolships' in this section; they are basically big, shallow copper swimming pools that the wort hangs out in to cool, which also allows the natural yeasts and bacteria in the air to settle on the beer and start what's known as the spontaneous fermentation process.

* apart from two with special permission

Orval

ABV
6.2%

Country of origin
Belgium

Try it if you like
Dry muscat or
farmhouse ciders

Great with
Roast pork belly or
mandarin cheesecake

Also try
Tilquin Oude Gueuze
à l'Ancienne
6.4%, Belgium

Like Madonna, Pelé and Picasso, Orval only needs one word to reduce its fans to misty-eyed, wobbly-kneed messes – and I don't mean from the alcoholic strength of it either.

Orval is one of those unicorn beers that make people say 'I don't like Brett beers but...' or 'I'm normally a wine or cider drinker but...'. And even the most hardened, snobby sommelier has been known to say 'I don't like beer at all but...'. It's just that unique.

A certified Trappist beer, it's brewed at the sensationally beautiful Orval brewery in the Ardennes region of Belgium. The key to this beer is how it changes as it ages, a cause of much debate amongst the seriously nerdy corner of the beer world.

If you are more of a dry, farmyard cider drinker, then the older versions are for you, but if you are looking to graduate from fresher, bitter beers like IPAs, then younger, orangey, hop-forward versions are a good place to start. You can find the bottling date on the label to help you out with your choice.

Brasserie De La Senne Bruxellensis

ABV
6.5%

Country of origin
Belgium

Try it if you like
Fruity, leathery whiskies

Great with
Confit duck leg

Also try
Boon Oude Geuze
Mariage Parfait
8%, Belgium

I may have mentioned earlier that I really like this brewery, so it's a rare second appearance for anyone in here, because it's just that good!

Bruxellensis is an unsurprising move for the pair, with Brussels being the only place where you can make a beer and call it 'Lambic', due to the unique yeast and bacteria in the air. However, the inoculation of this beer happens in the bottle instead of in big coolships.

The flavours are intense and too many to list here in full, but expect full-on red berry fruit on the nose, with a hint of pear, a dry-as-a-bone leathery finish and a vibrant earthy pepperiness.

Brewery Verhaeghe Vichte Duchesse De Bourgogne

ABV
6%

Country of origin
Belgium

Try it if you like
Balsamic vinegar

Great with
Rib eye steak

Also try
Russian River
Consecration
10%, USA

I don't think I'm exaggerating when I say, that it's possible this – oud bruin – is one of the most divisive beer styles in the world, but I love it with a passion that will not be dimmed or dismissed!

Brewery Verhaeghe Vichte is a small family-owned brewery in southwest West Flanders, Belgium. The brewery dates back to the late 1800s and, while it may not have quite the global reach of its nearby neighbour, the famous Flanders red beer producer Rodenbach, it is definitely, in beer geek circles, wildly revered.

The beer style is one of the only that I know of that celebrates the presence of acetic acid with such flamboyance. Acetic acid is what creates vinegar, and can be the line in the sand when it comes to this style.

While acetobacter has been shown to be present in the Flanders red style, it just doesn't shine through as aggressively as in the oud bruin style, that Duchesse flaunts with such abandon.

Leathery, balsamic, cherry, bitter dark chocolate, and a hint of Pedro Ximénez sherry, it is a wildly complex beer that is so utterly perfectly with an enormous rib eye steak.

Lost Abbey
Duck Duck Gooze

ABV
7%

Country of origin
USA

Try it if you like
Fino sherry

Great with
Oysters

Also try
Grutte Pier
Tripel Uit De Ton
8%, Netherlands

Although I don't want to put lots of rare or difficult-to-find beers in here, it's hard to avoid in this section as these beers take so long to make and the yield is so low.

Much as it would be tempting to make a joke about only choosing this one for its name (it is, after all, a most excellent pun), I haven't – it's all about the beer.

In fact, if you get into your complex then just keep an eye out for Lost Abbey beers full stop. Brewmaster Tomme Arthur sure knows what he's doing.

Duck Duck Gooze is a homage to the gueuze style of beer from Belgium and, just like those beers, it's a blend of old and young beers from different barrels.

The initial impact is like drinking white balsamic from a leather gourd (I mean that in a good way, odd as it sounds), but it's soon joined by some wood complexity and flirty notes of sea buckthorn. It's a palate awakener extraordinaire.

Collective Arts Brewing Jam Up The Mash

ABV
5.2%

Country of origin
Canada

Try it if you like
Tasteful beers and cans

Great with
Montreal smoked meat

Also try
Bellwood's Dry Hopped Gose
4%, Canada

Collective Arts doesn't just make beer, they also make spirits and pre-mixed cocktails. They aim to fuse the creativity that they put into those products (which are seriously delicious – honestly, not had a bad drink of any kind from these folks yet) with limited-edition works of art on the labels.

Promoting diverse artists from different walks of life is very much part of the ethos and the cans look as pretty as the liquid inside them tastes.

The first time I had this beer, I was sure there was fruit in it, it is that full of zingy flavours but it turns out it is all done through dry hopping – it's a bit like an acidic will-o'-wisp zipping around your mouth. I almost thought that in all the pretty art, they'd failed to mention it, which just goes to show even the professionals get it wrong some times!

Whiplash Suckerpin

ABV
3.2%

Country of origin
Ireland

Try it if you like
Fizzy sweets

Great with
Mackerel

Also try
Smokey beer styles called Grodziskie, not often brewed but interesting

I'm not saying the lads who own Whiplash are a bit into beer or anything, but it began life as a 'fun side project' for Alan Wolfe and Alex Lawes in 2016, when they were both already working in brewing.

Thing is, they're really good at what they do so by 2017, in their words, 'Whiplash was its very own pain in the neck' and by 2019, they'd opened their own brewery in Ballyfermot, Dublin.

The artwork on all the cans is really quite something, created by Irish artist Sophie de Vere. I've literally seen people stopped in their tracks by it and others just buy a can without even looking at what's inside.

If they bought this beer though, they'd better pucker up for a heck of a palate ride! It's dry-hopped Berliner Weisse, with sparky tartness from the brewery's house-cultured lactobacillius. It's fermented on peppery Belgian yeast and smothered in Lemondrop hops, which bring exactly the flavours you'd expect from the name.

Original Ritterguts Gose

ABV
4.7%

Country of origin
Germany

Try it if you like
Sherbet

Great with
Fried chicken

Also try
10 Barrel Brewing
Cucumber Crush
5%, USA

There's a lot to be said for the theory that the beer revolution has been driven by the curiosity of home brewers and their desire to dig deep and often, if many of my friends are anything to go by, out-nerd each other!

In this case, it was a German home brewer, Tilo Janichen, who decided that it was vital he resurrect the style of his local area, Leipzig, and set about finding breweries to work with to do so.

Most of the Goses you will find these days have fruit added to them, but this is brewed in the traditional style, with coriander seeds and sea salt, and is soured naturally with Lactobacillus, making it very tart, refreshing, light, savoury and a little spicy.

Sour Beer Scallop Ceviche

<u>Serves 4</u>

- 8 large, plump scallops
- 300 ml (10 fl oz/1¼ cups) Berliner Weisse (passionfruit or citrus for preference)
- ½ teaspoon fine sea salt
- ¼ teaspoon very, very finely chopped red chilli
- micro herbs, radish or salad cress, to garnish

Ceviche is simplicity itself, but it takes some knowledge and care to make it a success. The first thing is to make sure you get your scallops from a reputable source and that they're incredibly fresh. Fresh scallops should be firm, opalescent and can come with or without roe.

If yours have roe, then don't waste it, it's delicious. Reserve it and fry it incredibly quickly in butter and place in the middle of your circle of ceviche for an added touch of luxury.

Remove any muscle from the side of the scallop and place in the freezer for 15 minutes.

Mix together the beer and salt until the salt is dissolved, keep cold.

Slice each scallop as thinly as you can and lay in a single layer in a shallow, non-metallic dish. Pour over the beer mix. Leave for 3–5 minutes, or until the scallops start to turn opaque.

Gently lift the scallops out of the beer, portion the slices out evenly and arrange in a circle on small plates, sprinkle the chilli evenly across all the plates and garnish with your cress or micro herbs.

Ninkasi's Kiss

Serves 2

- 30 ml (1 fl oz/⅛ cup)
  Noilly Prat vermouth
- 50 ml (1¾ fl oz/¼ cup)
  elderflower liqueur
  (I use Chase)
- 1 egg white
- 2 teaspoons simple syrup*
- ½ teaspoon yuzu juice
- 2 ice cubes
- 200 ml (7 fl oz) Orval
  (or Wild Beer Co. Ninkasi)
- a pinch of sumac, to finish

Named for the Sumerian goddess of beer (and because I used Wild Beer Co.'s Ninkasi in the original recipe, which I also recommend) this cocktail can add a little kiss of sophistication to your beer-drinking repertoire.

Put the vermouth, elderflower liqueur, egg white, syrup and yuzu juice in a shaker with the ice cubes, shake for 20 seconds and strain into 2 flutes.

Top carefully with ale as it will fizz up (but it does look good with a bit of a head of both beer foam and egg emulsion on it). Sprinkle with sumac and serve.

*Simple syrup is equal parts sugar and water warmed gently until the sugar dissolves and then cooled again.

7

The
Dark
Side

Dark beers get a really bad rep, and they really don't deserve it. People seem to think they are heavier, more calorific or all taste the same – basically people equate dark beer with that ubiquitous Irish brand that you see all over the world, without ever having tried something like, say, Moorhouse's Black Cat Mild, which is as light and refreshing as they come.

Now I'm not saying that Guinness is particularly bad, but its serving method using nitrogen, which has very tiny bubbles and makes the beer seem thicker and more viscous, has a lot to do with this misconception that all beers are the same, they aren't!

So, I'd like you to think about these beers differently and don't just drink with your eyes. Do you like chocolate, coffee, dark fruits and fruit cakes? If you like one or all of those things, then I can almost guarantee that you will like some or all of the beers in this section.

Oh, and before you go, I just want to bust a myth that just won't go away... there isn't any iron in dark beers (or any beers for that matter) but all beer does contain niacin, which is a vital precursor for the uptake of iron in the blood. Another good pub quiz fact!

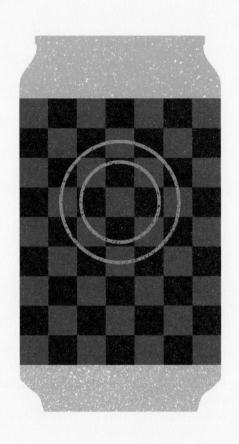

Baxbier
Koud Vuur

ABV
6.3%

Country of origin
Netherlands

Try it if you like
Smoky whiskies

Great with
Braised beef ribs

Also try
Durham Brewery
Temptation
10%, UK

Baxbier is from one of my favourite places in the world, Groningen. It may seem like an odd place to love, but it holds one of the best independent beer festivals with the warmest welcomes ever.

Baxbier is a success story I have kept an eye on from the start. They've gone from just being stocked locally to being sought after all over the country, and word is getting out about them internationally too.

The Koud Vuur is, for me, their most accomplished beer – delicately smoky, full of blackcurrant and bramble fruits and with a thirst-slaking dry finish.

These young men have a serious future ahead of them, which is just as well as they've built a pretty big brewery on the outskirts of the town. However, I have faith that they will outgrow even that in the next few years.

Odell
Cutthroat Porter

ABV
5%

Country of origin
USA

Try it if you like
Stouts

Great with
Raclette cheese

Also try
Three Boys
Oyster Stout
6.5%, New Zealand

Doug Odell is one of my heroes in the brewing world, and I've never met anyone in the industry with a bad word to say about him and whilst he may have taken a step back from the brewery, he is still active in the beer world and force for good, having ensured the independent future of the business by making it employee-owned.

In a portfolio of beers with exceptional balance and elegance, this is still, despite me changing my mind a hundred times, my favourite beer from his brewery.

Named after the Coloradan state fish (yes, there really is such a thing!), it's like someone made a fruit cake, added some cocoa and then smothered it in blackcurrant jam – a description I've used before, but one I can't seem to better when describing this beer.

Designed to sit somewhere in between a porter and a stout, it offers a lot of the richness of the latter and the drinkability of the former, and when I finally get the chance to grab a fly rod and go fishing for these notoriously fearsome fighting fish, I will make sure I have a bottle in my bag to celebrate or, more likely, commiserate at the end of the day.

Moorhouse's Black Cat Mild

ABV
3.4%

Country of origin
UK

Try it if you like
Coffee

Great with
Another

Also try
BoxCar Mild
3.6%, UK

This was my introduction to mild beer, and what a welcome introduction it was. Even if Black Cat started life as a bit of a hail Mary brew initially, because at the time the company's main focus was hop bitters (low-alcohol mild, stout and bitter) for export and use in shandy concentrates.

The beer was initially rustled up, with some concentrated liquorice that was lying around the brewery, because a local beer festival asked for a low-alcohol dark beer. When the beer sold out in double-quick time, they realised they had a potential hit on their hands.

The brewers started to experiment with more traditional ingredients, and when they included chocolate malt (a malt kilned at high temperatures) to provide the depth of colour and flavour, Black Cat as we now know it landed on the beer map on all four paws.

It has endured, even if its popularity has waxed and waned, and ownership has changed hands several times – the beer is still the same light mix of chocolate, coffee and dried fruits, providing humble refreshment without pretension.

La Sirène Praline

ABV
6%

Country of origin
Australia

Try it if you like
Praline chocolates

Great with
Raspberry ripple
ice cream

Also try
Cigar City
Marshal Zhukov's
Penultimate Push
11.5%, USA

In the UK, there is a woodstain called Ronseal whose slogan is 'Does exactly what it says on the tin'. This is what we would colloquially call a Ronseal beer.

Fascinated with their spontaneous fermentation beers as well as this one, I got to spend the afternoon at the brewery, myself and head brewer and co-founder Costa Nikias nerded out in the cold store until we were both shivering. Then I spent hours being plied with beer and cheese by co-founder James Brown, which led to me being more than a little outspoken during my appearance on the Ale of a Time podcast, in front of a live audience, during Melbourne's Good Beer Week a bit later on... anyway!

Although they specialise in farmhouse ales, it's this one that really took my breath away, which is saying something considering how good their Saisons are. Using a Belgian stout as its base (a dark beer using a fruity Belgian yeast), it's made with what are basically the ingredients for Nutella®, and that's exactly what it tastes like. The sweet, unctuous body comes from the use of both lactose and hazelnuts, meaning this beer punches well above its 6% ABV weight.

Rochefort 10

ABV
11.3%

Country of origin
Belgium

Try it if you like
Pedro Ximénez

Great with
Prune cake *(Far Breton)*

Also try
St Austell
Black Square
10.6%, UK

I'm not known for entering a room quietly, but even I was amazed at the stir I caused when I visited Rochefort. I went with a couple of other beer writers and there was a foot of snow on the ground and a fairly unpleasant, sleeting storm outside as we walked across the enormous grounds of the abbey to the brewhouse. It was only when I felt warm enough to take my hood down and my scarf off my face that the 'mistake' was discovered.

According to the Prior who was conducting our tour, I was the first woman to be allowed – albeit accidentally – into all areas of the brewery. I have no idea whether it's true or not, but I'm not about to start calling senior monks fibbers!

Rochefort 10 is, quite simply, a god-given nectar. Prunes, raisins, currants, dried cranberries, apricots, cinnamon, nutmeg, orange peel, brandy, Pedro Ximénez sherry and so much more abound in this beer. A warning though – it's dangerously strong, so do be careful if you're going in for a second, or at least make sure you're sitting down!

Baltika
No.6 Porter

ABV
7%

Country of origin
Russia

Try it if you like
Coffee chocolates

Great with
Smoked salmon

Also try
Põhjala Öö
10.5%, Estonia

I bet you weren't expecting something called porter to turn up in this book, but it's a testament to how beer styles can change so fundamentally when they move across borders.

This is a style that has much myth and legend about it. To put it simply, porters were exported from UK breweries to the Baltic states a lot in the 19th and early 20th centuries. When successive wars dried up that supply, local breweries took up the mantle. With all the tradition being more based around lagered beers, this was the type of yeast used and that's how this style came about. This may upset the purist beer historians because of its simplification, but it's the easiest way I can explain it in brief. If you want to know more, I strongly recommend reading the works of Ron Pattinson and Martyn Cornell.

I have literally lost count of the amount of awards this beer has won in the Baltic porter category because it absolutely nails the style.

Coffee, dark roast and a lashing of liquorice with an overtone of astringent redcurrant, it's a complex and fascinating beer that you have a tendency to look at faintly confused when you take your first ever sip, before diving back in for more.

The Kernel
Export India Porter

ABV
6%

Country of origin
UK

Try it if you like
Espresso

Great with
Honking blue cheese

Also try
Deschutes
Black Butt Porter
5.5%, USA

I'm ashamed to say that the first time Evin O'Riordan handed me one of his beers to try, I feigned deep enthusiasm and put it in my fridge for six weeks before my acute embarrassment at seeing this incredibly nice man in the distance at the pub, or around London's famous Borough Market, led me to crack it open. And boy was I impressed at what a great beer it was, and immediately rushed to tell him so.

To which, in his usual laconic style, he responded: 'Well that's good, I've just put a deposit on a brewery.'

Now, I feel I need to put some context to this. At the time, I had just had a run of enthusiastic folks with ambitions of turning pro pressing truly terrible home brew into my hands, and I had not only bad beer fatigue, but the sight of their faces as my lack of filter between brain and mouth crushed their dreams was something I simply couldn't face doing to an acquaintance who I was fond of... turns out I worried over nothing.

The Kernel has been a reflection of O'Riordan's personality from the start: softly-spoken with its brown paper bag branding and wonky typed labels, a steely determination to carve its own path and having little or no truck with trends or hype, despite frequently being the creator of them.

And this is because O'Riordan is the true beer nerd's nerd. He is very much in it for the purity of the product without being witless about it. While not often-sighted (he doesn't really do festivals or the like due to being really rather shy), if he is out and about he'll be having a quiet conversation in a corner with a like-minded soul. Perhaps it's why he employs some of the bubbliest faces in the beer scene and brews ales that speak for themselves. Whatever the truth of it all, The Kernel is one of the cornerstones of London's, nay the UK's, craft beer scene and long may that continue.

The Export Indian Porter's ABV really should have an '-ish' after the ABV, because one of the both brilliant and (as a writer) infuriating things about The Kernel beers is that they vary slightly from batch to batch. Different hops, slightly different strengths, slightly different characters... they're a bit like identical twins: when you see them apart it's hard to discern the differences from one to the other by memory alone, but when together you can tell immediately.

Regardless of the hops or the ABV, the beer itself is based on an old Barclay Perkins recipe, and always has that biting hop character in some way and almost always a floral hop, all of which is underlined by a complex cocoa and coffee malt base with a tiny whisper of smokiness and some juicy hedgerow fruit too.

J.W. Lees
Harvest Ale

ABV
11.5%

Country of origin
UK

Try it if you like
Tawny Port

Great with
Stilton cheese

Also try
Minoh
Imperial Stout
8.5%, Japan

I love it when a beer has a great story behind it, especially one with a bit of pride, and this one has exactly that. During a brewer's dinner in the mid-80s, everyone was chuntering about the rampant takeover of lager (or 'eurofizz' as it was rather childishly nicknamed) and so the then head brewer of Lees decided that he would showcase the very finest of the UK's barley and hop crop and make a harvest ale.

A nod to the UK tradition of churches and schools getting together to celebrate and give thanks for the gathering of crops, it uses East Kent Goldings and British barley and that's it. Its heady 11.5% alcohol content means that when it's fresh, it's redolent of fruit cake, maple syrup and orange rind but, as it ages, it becomes more oxidised and like fortified wine.

In fact, some of the older bottles I have opened have so many Madeira, Port and sherry characteristics that it is almost impossible to think of them in any other terms than beers for a special occasion – which can just be a Tuesday because waiting for that perfect moment, means it may never arrive.

Good George Rocky Road

ABV
5%

Country of origin
New Zealand

Try it if you like
Rocky road ice cream

Great with
Child-like glee

Also try
Brouwerij Martinus
Smoked Porter
9%, Netherlands

There's nothing like a beer that makes you giggle, even though it's the first one of the day. That's what this beer did for me when I first tried it as I settled into the local beer café on the shores of Lake Taupo in New Zealand.

Another 'Ronseal' beer, it does quite literally what it says on the tin – and in a tin it does indeed come. Modern cans are great for beer, as they prevent damage from light and oxygen and could be more environmentally friendly than bottles.

Rocky Road is the result of Good George joining forces with local chocolatiers Donovan's to make some seriously good-time beers. The flavours are individually clear as a bell – raspberry, chocolate and vanilla – but they also harmonise to give you that feeling of drinking a bowl of booze-spiked rocky road. I applaud any brewery that can achieve that technical feat so flawlessly and yet still talk about it with childish glee.

Fyne Ales
Vital Spark

ABV
4.4%

Country of origin
UK

Try it if you like
Dark lager

Great with
Mature Cheddar cheese

Also try
Birrificio del
Ducato Verdi
8.2%, Italy

Mild beer is a matter of much historic debate, but I will always bow to the beer historian Martyn Cornell on the definition. He says they were thus-named as they were 'fresh' beers, as opposed to having some age on them, and, as a result, had less hop character. However, they were by no means unhopped or low alcohol as a lot of people seem to want to rewrite them as.

Fyne Ales' version in Vital Spark certainly brings the word 'fresh' to mind when you drink it, because it is just that. Lean of body and hopped with the tropical Amarillo and grapefruity Cascade, both being applied sparingly. I almost wish that the brewery, a genuine gem hiding up in Argyll in Scotland, was somewhere more metropolitan so they could get all the plaudits that I think they deserve, but then perhaps they wouldn't have the quiet, contemplative approach to great beer making if they were. Either way, grab anything of theirs you see. You won't regret it.

Quick Chocolate Pots with Kriek Cranberries

Serves 6 as a small treat
or 4 indulgently

- 200 g (7 oz) dark chocolate
  with at least 70% cocoa
  solids, broken into pieces
- 100 ml (3½ fl oz) spiced
  or barrel-aged stout
- 100 ml (3½ fl oz/scant
  ½ cup) double (heavy) cream
- 2 teaspoons regular
  olive oil

For the cranberries:

- 25 g (1 oz/¼ cup)
  dried cranberries
- 50 ml (1¼ oz/¼ cup) kriek
  (see beer list below)
- 2 teaspoons caster
  (superfine) sugar
- tiny pinch of sea salt

Beer list:

- Oud Beersel Kriek - Belgium
- Odell Friek - USA
- Boon Kriek - Belgium
- New Glarus Wisconsin
  Belgian Red - USA
- Russian River
  Supplication - USA

There are times when you just feel like you deserve a treat and these little pots are perfect. I had the idea when Tempest Brewing Company sent me some of its bourbon barrel-aged Mexicake – a kind of mole stout, and that's what I used when I made this, but there are plenty of other examples like it out there or you can substitute a barrel-aged stout – bourbon barrel ones work well.

Prepare the cranberries first. Put all the ingredients in a small saucepan and warm over a low heat until the liquid just starts bubbling. Stir well, then allow to bubble for 4–6 minutes until the liquid is almost gone. Leave to cool.

To make the chocolate pots, melt the chocolate, using a microwave or a heatproof bowl set over a pan of simmering water. While that is melting, put the beer in a small pan and heat until it just starts to bubble around the edges. Swirl the pan a few times but don't let the beer boil.

When the chocolate has melted, take it off the heat and add a small amount of the hot beer, beating with a rubber spatula until the mixture is smooth – don't panic if it looks granular, you just need to apply some elbow grease! Repeat until all the beer is used and you have a smooth, shiny paste.

Add the cream and oil and beat until you get a smooth mixture. Pour into espresso cups or ramekins (custard cups), size depends on whether you're making four or six servings. Place in the fridge for 30 minutes to set.

Take out of the fridge, top with the soaked cranberries and serve.

Peat Nut Monster

Serves 2
(in tumbler-style glasses)

2 ice cubes
100 ml (3½ fl oz/scant ½ cup) peated whisky
1 handful of peanut butter M&Ms® or Reese's Pieces®
500 ml (17 fl oz/2 cups) Imperial Stout, at room temperature

My other half thinks I'm crackers for liking this cocktail. In fact, he looked at me with such utter horror when I made it that I genuinely considered not putting it in here. But I really liked it so, nerrrrrr.

Let me know whether you think I'm an absolute monster or a genius for this via social media – but only with a picture of the cocktail. If you haven't tried it, you don't get a say!

Blitz the ingredients in a blender until totally combined. Pass through a fine sieve into a shaker, top up with beer and stir gently. Pour into two tumblers to serve.

Fruity Numbers

First things first, fruit beers don't have to be sweet – in fact I have a confession to make here, I don't have much of a sweet tooth (unless it's pick 'n mix and then I shouldn't be allowed to handle one of those scoops unattended).

So this chapter is a real mixture; there are some traditional fruit beers that span back over centuries, there are some great fun beers where the brewer has let loose their inner child in the candy store and there are some serious beers too (but not too serious because, after all, that's not what beer is about).

All in all I'd like you to put aside any potential prejudices against fruit beers and give them another go (this is also where I wish that there was a pea beer, so I could ask you to give peas a chance, but no one has made one yet, so I'll just have to keep that terrible dad joke for another day).

Oud Beersel
Oude Kriek

ABV
6.5%

Country of origin
Belgium

Try it if you like
Cherry pie

Great with
Peking duck

Also try
Funkwerks
Raspberry Provision
4.2%, USA

Choosing a traditional Belgian brewer for this section was incredibly tough, so I'm going to start this bit by saying if you like this try more – try Tilquin, Boon, Duchesse de Bourgogne, 3 Fonteinen, Lindemans, Cantillon, Petrus and Rodenbach – because these are the guardians of a unique and wonderful style of beer.

Anyway, back to Oud Beersel. Founded in 1882, the brewery fell into financial troubles in 2002, but in 2005 it was rescued by two friends, Gert Christiaens and Roland De Bus. Inheriting an aluminium brew kit (not advisable at all), they had to outsource the actual making of the beer to Frank Boon.

The good news is that there will soon be a brewery back up and running at Oud Beersel and the beers, which have remained barrel-aged and blended on site, are sensational. This is my favourite of them all – brisk, complex, leathery, almondy, like a tart cherry pie, which can be explained by the fact that there is at least 400 g (14 oz) of cherries used for each litre (34 fl oz) of beer. The brewery has plans to plant an orchard so that they can source their cherries as locally as possible.

Against The Grain Bloody Show

ABV
5.5%

Country of origin
USA

Try it if you like
Blood orange gin

Great with
Kentucky-style
fried chicken

Also try
Hitachino Nest
Yuzu Lager
5.6%, Japan

You know when you meet people and you can see EXACTLY why they've called their business something like 'Against the Grain'? Well, these chaps are certainly living their brand.

Founded by Jerry Gnagy, Sam Cruz, Adam Watson and Andrew Ott, this brewery is a collection of the mad, bad and dangerous to know.

A true success story, the brewery started as a small operation with an attached smokehouse in 2011. Now with a production brewery just a few blocks away, their beers are found in 43 states and 25 countries.

The Bloody Show is clearly designed for those hot, sultry Kentucky days, when the air barely moves and you can almost feel an attack of the vapours coming on.

Originally a collaboration with Danish brewery Mikkeller, it's packed full of orange peel and blood orange purée and lots of fresh complementary hops like Mosaic and Huell Melon. But, drinker beware, it may taste like the poshest shandy you've ever drunk, but it does pack a pithy punch.

8 Wired
Feijoa

ABV
Various

Country of origin
New Zealand

Try it if you like
Bazooka Joe bubblegum

Great with
A confused expression

Also try
Pfriem
Oude Kriek
5.5%, USA

If you are ever in New Zealand, don't ask what a Feijoa is in front of a group of natives. If my experience of doing so at 8 Wired's brewery is anything to go by, it's like saying you don't know that the sky is blue or water is wet.

Don't ask what Feijoa tastes like, either. No single person will give you the same answer, but 20 different folk will spend a good 10 minutes giving you their version. This must be because they don't sell Bazooka Joe bubblegum down there and that's EXACTLY what it tastes like (well, to me anyway).

This beer is a glorious romp through the culinary unknown for most people, but it's a fun one nonetheless. The beer is brewed as a fairly simple pale ale, then poured into different barrels, which are impregnated with various wild yeasts and bacteria and left to age for about a year before the lunatics at 8 Wired shove 800 kg (1750 lb) of fruit through the barrels' bung holes (around 30 in total). They are then sealed up for around another year to pick up all the fruit's flavour.

The end result is like sour gumballs in beer form, and who couldn't be happy with that?

Dogfish Head SeaQuench

ABV
4.9%

Country of origin
USA

Try it if you like
Margaritas

Great with
Oysters

Also try
Anderson Valley Brewing
Briney Melon Gose
4.2%, USA

Dogfish Head is one of the longstanding stalwarts of the US craft brewing scene. Founded by Sam and Mariah Calagione in 1995 in Reheboth, Delaware, at the time, it was the smallest commercial brewery in America... Well how things change.

The brewery grew in stature and popularity year-on-year, producing fascinating beers like Palo Santo Marron, which is aged in huge vats of the wood made from the same name, and occasionally undertaking crazy projects like making chicha (page 147). They always seemed one step away from biting off more than they could chew, but they always managed to pull it off.

In 2019 they 'merged', as they put it, with the USA's largest independent brewery, Sam Adams, and have continued to brew their 'off-centred beers for off-centred people'.

SeaQuench is absolutely stunning. I can't put the beer into better words than the brewery does: 'a session sour mash-up of a crisp Kölsch, a salty Gose and a tart Berliner Weisse brewed in sequence with black limes, sour lime juice and sea salt. The result? A citrusy-tart union that has captured the attention and hearts of beer, wine and margarita drinkers alike!

Stiegl
Radler Grapefruit

ABV
2%

Country of origin
Austria

Try it if you like
Fruity sodas

Great with
A long bike ride

Also try
Rothaus
Radler Zäpfle
2.4%, Germany

Just think about it: you've decided that you are going to squeeze yourself into some Lycra® and head off for a spin; maybe you think you can take on a mountain stage in Austria? If you're not gasping for a beer at the very thought, then I'm afraid we might not be friends.

But, if you are, then a Radler is your go-to buddy here. Most of the traditional ones are based on lagers or wheat beer. Notable versions include Germany's Schöfferhofer and 'craft' versions like the UK's Marble Sunshine Radler based on a simple wheat beer, but all of them are like nectar from the gods after a serious bout of exercise.

Originally designed to refresh cyclists without high alcohol levels, this is one of my particular favourites. In fact, Stiegl is just a good all-round brewery and the first in Austria to practise an organic approach. (Stiegl is a brewery with a long and fascinating history, including links with Mozart!)

Pithy, bright, grapefruit flavours, middling carbonation and without anything cloying on the back of the throat, it slips down in big gulps that are nearly always followed by a satisfied 'ahhhh' at the end.

Flying Monkeys
12 Minutes to Destiny

ABV
4.1%

Country of origin
Canada

Try it if you like
Rosé wine

Great with
Tuna nigiri

Also try
Surly Brewing Rosé
5.2%, USA

When an idea is right, it's right. The name of this beer tells you everything you need to know about how long it took to nail this delightfully fresh and fragrant lager.

Founded by Peter Chiodo, the brewery loves to hire passionate and talented home brewers and its motto is 'brew fearlessly', which is to be applauded in an age when big beer marketing budgets could possibly take you out in a heartbeat.

Based in Ontario, the taproom knows how to welcome people in, with a vast array of pinball machines and classic arcade games as well as great food on offer. So you can settle in and enjoy your stay.

This particular beer is made with hibiscus flowers, rose hips, fresh raspberries and orange peel; as you can imagine, it's a complete riot of fruity, floral flavours and aromas, with a good bop of citrus bitterness at the end... in all honesty, it's simply summer in a glass.

Amundsen
Lush Passion Fruit

ABV
5.2%

Country of origin
Norway

Try it if you like
Summery flavours

Great with
Cantonese-style duck

Also try
Wiper and True
Purple Rain
4.8%, UK

Amundsen is probably Norway's best-known craft brewery, and for good reason.

Innovative beers, stunning branding and an ability to grab the zeitgeist, are things I think they really excel at and they've become popular all over the world as a result.

Their 'Dessert in a Can' range is ridiculously fun – taking classic desserts and making them in beer form – and their innovation in the pastry sour department is also great.

But it's the Lush I return to, a raspberry and lime Berliner Weisse. It reminds me of the ice pop lollies you used to get when you were a kid, and anything that brings back happy memories like that deserves all the plaudits it gets.

Watermelon, Mint and Chilli Pickle

Serves 4

- 15 g (½ oz/1 tablespoon)
  fine sea salt
- 15 g (½ oz/1 tablespoon)
  caster (superfine) sugar
- 1 teaspoon chilli
  (hot pepper) flakes
  (optional)
- 20 ml (¾ fl oz/generous
  1 tablespoon) hot water
- 330 ml (11¼ fl oz/1⅓ cups)
  well-chilled cucumber
  or watermelon sour beer*
- 50 ml (1¾ fl oz/¼ cup) rice
  wine vinegar
- ½ small watermelon, chopped
  into roughly 5 cm (2 in)
  square-ish chunks
- cold water (optional),
  to top up
- 10 mint leaves, chopped,
  to garnish

*If you can't find a cucumber
or watermelon beer, use a Berliner
Weisse or Gose like Ritterguts
Original Berliner Weisse or
Magic Rock Salty Kiss.

I love this with fried chicken. It's a little more robust than most quick pickles so you can make it about 2 hours ahead of eating.

Put the salt, sugar and chilli flakes, if using, in a large bowl and add the hot water. Stir vigorously to dissolve the salt and sugar.

Add the beer and rice wine vinegar, and then the watermelon pieces and, if needed, enough cold water to cover the melon. Refrigerate for no longer than 2 hours.

When ready to serve, drain the melon and sprinkle with mint.

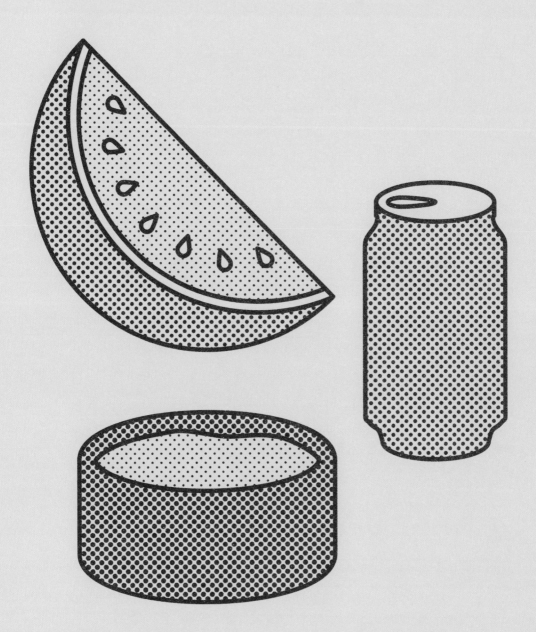

Rube Icon

Serves 1

- 50 ml (1¾ fl oz/¼ cup)
  lychee liqueur
- 2 teaspoons lime juice
- 150 ml (5 fl oz/⅔ cup)
  tropical fruit-flavoured
  beer (preferably sour,
  or add 1 tablespoon
  triple sec)

Another terrible play on words, this is a boozy version of the tropical drink Rubicon, which was a feature of my childhood. I'm sure all over the world there are tropical punch-flavoured drinks that you'll identify as similar.

Shake the lychee liqueur and the lime juice together over ice and strain into a martini glass. Top up with beer, stir gently and serve.

9

No,
Low and
G-Free
Beers

Believe it or not, there are times when I really, really wish there was a decent low- or no-alcohol beer on hand. It might be one of the rare occasions when I'm driving and I actually just don't want to drink sweet, fizzy nonsense all evening, but options are all too often few and far between.

And, while not being gluten intolerant or suffering from coeliac disease myself, my heart goes out to those who can't have gluten and may feel they are deprived of beer.

So, while I think that genuine high-quality beers of this ilk are still too difficult to get your hands on, there are a number spread over the world that I think are well worth drinking for aesthetic reasons as much as any other.

It's worth noting that, in most places in the world, anything below 0.5% is considered non-alcoholic and is physically impossible to get drunk on, so please don't think when it says 0.5% that I'm being dim (although anyone who knows my legendary lack of numeric capabilities could easily believe otherwise!).

Big Drop
Uptown Craft Lager
(GF/AF)

ABV
0.5%

Country of origin
UK

Try it if you like
Easy drinking lager

Great with
Halloumi salad

Also try
Pistonhead
Flat Tire Alcohol-Free
0.5%, Sweden

When founder Rob Fink realised that his legal practice was leading him to drink more than he'd like, he cast around for genuinely enjoyable non-alcoholic beers and came up pretty blank. So he decided to do something about it. He hired a brewer who knew what he was doing in Johnny Clayton and the rest, as they say, is history.

Using a 'lazy' yeast that produces very little alcohol, the company has now gone from strength to strength. It produces what are widely recognised as some of the best low-/no-alcohol beers in the UK, and is set to conquer the world if Rob has his way.

The lager is one of the few in the range that doesn't contain lactose (milk sugars), so is also vegan-friendly as well as being gluten-free. Plus, it certainly hits the nail on the head of that refreshing hit you want from a lager at the end of a busy day or a sweaty workout.

Light amber and with a slight fresh hay and subtle caramel nose, it finishes with a pleasant pepperiness that really lights the whole affair up. It's a welcome change from the bland and boring low-alcohol lagers of the past.

Stiegl Freibier
(AF)

ABV
0.5%

Country of origin
Austria

Try it if you like
Skiing in a straight line

Great with
Weinerschnitzel
(Viennese schnitzel)

Also try
Athletic Brewing
Free Wave IPA
0.5%, USA

There really aren't a lot of breweries that are still in existence and can claim that they've been visited by Wolfgang Amadeus Mozart, but Stiegl can. Being the sad case that I am, I've actually sat with one of this beer's full-blooded cousins and listened to a Mozart flute concerto and I can attest that it's a very pleasant experience.

My long-lapsed love for that particular woodwind instrument aside, this really is the real deal when it comes to excellent alcohol-free beer. It doesn't really surprise me because there is little Stiegl puts a foot wrong on.

As with most of their other beers, the ingredients are all sourced from Austria. Unfiltered to ensure maximum flavour, it has an unusually spicy tangerine pop on the nose and palate from the local Saphir hops.

Why Craft No/Lows Taste Better These Days

Ok, if you tried Kaliber back in the day, you're probably having a small coronary at the idea of a no-/low-alcohol section in a book like this. First of all, please breathe!

A lot of the problem with alcohol-free and very low-alcohol beer has been down to the different ways that they can be made.

Up until recently, the main way has been to boil off the alcohol, which effectively also just drives off nearly all hop aroma characteristics and almost returns the beer to its wort-like flavour, but with an odd bitter note left over.

Basically it has a tendency to taste like sweet, slightly bitter, boiled veg water – mmmmmmmm!

An alcohol extraction plant that works at very low temperatures is probably one of the best ways to make consistently tasty no/low beers (which is what Adnams has chucked a serious chunk of cash at recently and boy does it show – the Ghost Ship is almost indiscernible from its full-blooded counterpart).

And then, finally, there's the less spendy way of making full-flavoured beers, which involves using some of the newly developed 'lazy' yeasts. These yeasts aren't interested in the primary sugar in barley, maltose, so they don't bother fermenting it out.

Meaning you can make very tasty beers that don't have anything taken away in the making of them. So please, give them another try, you might be pleasantly surprised!

Sobah Lemon Aspen Pilsner (AF)

ABV
0.5%

Country of origin
Australia

Try it if you like
Real lemonade

Great with
Shellfish

Also try
Kingfisher Radler
Ginger & Lime
0%, India

There's lots to celebrate about the Sobah brand. Owned and led by Aboriginals, it's the first alcohol-free dedicated brewery in Australia and believes in treading lightly and acting ethically.

As with all beers, it's designed to live in a social space, but the purpose of this enterprise extends far beyond that. The brewery also highlights some of the social issues that affect the indigenous population and breaks down a lot of the associated stereotypes; instead championing the positive contributions the Aboriginal and Torres Strait Islanders make to Australian society.

Rather charmingly calling itself 'bush tucker beer', the vegan-friendly Pilsner uses a native fruit – lemon aspen – which produces a bright, fresh, almost pink grapefruit flavour and aroma. (There's also a super-tasty finger lime Mexican-style lager in the range.) It is so unbelievably perfect for the hot weather you get in Australia that all I could think of was golden sands and surfing when I tried it.

Celia Dark
(GF)

ABV
5.7%

Country of origin
Czech Republic

Try it if you like
Root beer

Great with
Pork knuckle

Also try
Voll-Damm
7.2%, Spain

A dark, gluten-free lager is a rare beast indeed. In fact, if there is another commercially available one in the world, I haven't been able to find it (which is not to say there isn't, of course, I am knowledgeable but not all-seeing!).

Brewed in the beautiful historic Czech town of Žatec at the now Carlsberg-owned site, the brewery dates back to the 1700s and the lagering caves are still dug into the castle walls. (It's also home to some exceptionally good porter, but that's another story for another book!)

Deepest ruby brown, the beer has strong depths of coffee and chocolate, but with a fresh berry note and light body that allows it to be both complex and refreshing at the same time.

Westerham Brewery Co. Helles Belles
(GF)

ABV
4%

Country of origin
UK

Try it if you like
German lagers

Great with
Soft pretzels

Also try
Glorious Pilsner
4.5%, UK

Westerham Brewery Co. has been quietly making gluten-free beer for quite some time, but not really shouting about it. That's typical of the people who own it. They just get on with making really good beer that, for the most part, happens to be gluten-free, so I thought I'd shout for them.

Having worked in the city for a long time, owner Robert Wicks decided to jack it all in and become part of Westerham's proud brewing tradition. He even references some classic brews from the old Eagle Brewery that once stood very close to where his brand new brewhouse (with a rather lovely taproom) stands now.

Helles Belles, as the name suggests, is a Helles-style lager with all the easy-drinking character you'd expect. Brioche flavours abound in the excellently structured but light body, and it's redolent with herbal, spicy Hallertau Tradition hops, which you only need one sniff of to be immediately transported to Munich.

Wicklow Wolf Moonlight (AF)

ABV
0.5%

Country of origin
Ireland

Try it if you like
Not catching fish

Great with
A day's fly fishing

Also try
Guinness
Pure Brew
0.5%, Ireland

Wicklow Wolf's beers were some of the first I tried from Ireland that I was impressed with and they have gone from strength to strength ever since.

Again, another brewery that was influenced by the American beer scene, it was founded by two friends, Quincey Fennelly and Simon Lynch, who say it was an 'inevitability' that they would end up in the brewing business together.

With staff now comfortably in double figures, the brewery even has its own 10-acre hop farm – a rarity indeed in Ireland. It's flourishing under Lynch's watchful eye and no doubt benefits from his 20 years in the horticultural industry.

The beer itself is almost indistinguishable from a full-blooded American-style pale ale, redolent with grapefruit Citra hops, but with the added white grape note that comes from the European Hallertau Blanc hops too – a delightful sipping beer when idling away some hours on a bank, dangling a rod in the water. (And I speak from experience of dangling rods in the water, it's the catching of the damn fish that all too frequently eludes me... still, at least I always have a decent beer or two!)

O'Brien Lager
(GF)

ABV
3.5%

Country of origin
Australia

Try it if you like
No worries

Great with
A beach sunset

Also try
Omission Lager
4.6%, USA

Sometimes you just want to sit back and enjoy a seriously easy-drinking lager at the end of the day while watching the sun go down, and this O'Brien one ticks exactly that box.

I won't tell you it's anything that's going to rock your world, because sometimes you don't want that. You just want to drink a well-made lager that is going to ease your mind and soul and that isn't going to interfere with watching the world go by and this, my friends, is exactly that.

A simple, light, clean body with a hint of thyme and lemon, it will hit the spot at the end of a long week at work while keeping your head clear at just 3.5%.

Green's
Dry-Hopped Lager
(GF)

ABV
4%

Country of origin
Belgium

Try it if you like
Pale ales

Great with
Friends or a hot day

Also try
Leeds Brewery
Original OPA
0%, UK

Full of vibrant, zesty hop character and zippy, clean refreshment, Green's has come a very long way from its origins when, frankly, the beers were pretty average... although I'm sure to a beer-deprived coeliac they were a godsend!

Although a UK-owned company, the beers are brewed in Belgium at the only brewery that, back in the early 2000s, finally agreed to help Derek Green in his 20-year quest to make gluten-free beers after he was diagnosed and forced to cut out barley and wheat.

A chance meeting with a brewing professor whose daughter couldn't eat gluten either, and the dream was born. There's now a whole range, all of which I could recommend, but the dry-hopped lager is my favourite.

Glutenberg
American Pale Ale
(GF)

ABV
5.5%

Country of origin
Canada

Try it if you like
Fruit cocktail

Great with
Montreal smoked meat
or pastrami

Also try
Westerham Brewery Co.
Scotney Pale Ale
4%, UK

Glutenberg was the first of this brewery's beers that I tried and honestly thought, 'I like that', rather than 'That's OK for a gluten-free beer', and their flagship pale ale still continues to delight.

It has all the fresh, vibrant grapefruit and lime with a hint of pine that you'd expect from an American-style pale ale. This beer is perfect to keep in the fridge and drink while you contemplate dinner after a long, hard day at work. I am also a huge fan of the IPA these guys produce – it's very polished.

Mikkeller Drink'in the Sun (AF)

ABV
0.3%

Country of origin
Denmark

Try it if you like
Cloudy lemonade

Great with
Grilled white fish

Also try
Arcobräu Urfass
Alcohol-Free
0.5%, Germany

It's no big surprise that the daddy of cuckoo brewing, Mikkel Borg Bjergsø, was one of the first 'craft' brewers to come up with a low-alcohol beer full of flavour. Since he turns out ideas as fast as most of us breathe, it surprises me it took him as long as it did.

This American-style wheat beer is straightforward lemony, peachy refreshment at its finest and shouldn't be pressed into service with anything more complicated than a piece of grilled fish or chicken and a green salad – but then sometimes simple is all you need.

Erdinger Alkoholfrei (AF)

ABV
0.5%

Country of origin
Germany

Try it if you like
Running marathons

Great with
A sweat on

Also try
Fentimans
Lemon Shandy
0.5%, UK

I love that this is being marketed as an isotonic drink for people post-exercise. In fairness to Erdinger, there have been studies done that show that a low-alcohol beer consumed directly after exercising is very good for you and will help you recover quicker than anything except a sports drink. This alcohol-free version has the benefit of being lower in calories than most sports drinks, too.

It's a bit more muted than most German-style wheat beers, but that's not necessarily a bad thing if you find the banana and clove notes of the yeast overwhelming. Besides, ice-cold after a workout, any beer tastes good, and this has an extra healthy smug factor thrown in.

New Belgium Glütiny Golden Ale (GF)

ABV
5.2%

Country of origin
USA

Try it if you like
Lager

Great with
Tacos

Also try
Wold Top
Against the Grain
4.5%, UK

New Belgium is one of those places you dream of working; an employee-owned business with serious ethical, environmental and social credentials.

Just so you know, you'll see on the website or the bottle, some very carefully-worded blather about the fact that it's 'crafted to be reduced gluten' – which could easily be mistaken for marketing horse manure.

However, don't blame New Belgium or any other US brewery, as the FDA, unlike most other places in the world, refuses to allow any business to use the term 'gluten-free', one assumes because there are thousands of lawyers sitting around sharpening lawsuits about it.

Anyway, bureaucratic nonsense aside, this (and its sibling the Glütiny Pale Ale) is an intensely refreshing beer with gentle butteriness, and aroma from the lovely leathery Goldings hops and a bit more punch from the Cascade and Nugget.

Author Biography

Award-winning beer and food writer Melissa Cole's passion in life is getting people to learn as little or as much as they like about what she considers the finest social lubricant known to humankind.

Respected the world over for her fine palate, she is invited to judge at competitions in places as far flung as New Zealand, the US and Brazil, as well as closer to home across Europe. Aware of the impact this can have on the future of breweries, she has also spent years educating herself on the brewing process, often getting her hands dirty and making beers with some of the best brewers across the globe.

Melissa's other passion is food and she is recognised as probably the UK's leading expert in pairing and cooking with beer, having spent years dedicated to figuring out which beers work in which dishes and why – which she regularly calls 'making a whole series of disgusting mistakes, so you don't have to'.

This is Melissa's fifth book and keeps true to her passion for simple beer communication that neither baffles nor patronises the audience, she just wants people to enjoy beer as much as she does.

Thank You

Hardie Grant, once again, you are the best publishers anyone could wish to work with. Eila, thanks for your patience. Stuart Hardie, you've made it look fabulous; you are so talented and funny.

Ben, you deserve every book dedicated to you for putting up with me. Your support is boundless and your love endless. Thank you so much.

My family, mum and dad, Mel, Josh and Kate, you are my rocks.

Pam and Stan Eaton, this is still all your fault you know! And to Mike and Joanna Eaton, thank you for your love and for the wine breaks from beer.

My friends and beer family, I've missed you all so much this past 18 months. It's been very hard not to see all the faces I know and love, but we'll be back together soon.

There are just too many people to mention by name, so instead I want to mark the sad passing of two people who I will miss terribly and that leave a big hole, literally and metaphorically, in the beer and food worlds respectively.

Roger Ryman, who was head brewer at St Austell, was a friend, a mentor and a titan in beer, I didn't spend enough time with you but I'm proud we brewed Sayzon together.

And to Charles Campion, friend, eating pal, drinking buddy and genuine champion of diversity in food and of me. I hope one day to make radio contact with you again.

And finally, to all the people who stood up this year to try and weed out the worst of people in the beer world, and of course wider, I salute you.

Beer Index

Recipes and cocktails are in *italics*

The Ultimate Book of
Craft Beer

First published in 2021 by Hardie Grant Books,
an imprint of Hardie Grant Publishing

Hardie Grant Books (UK)
52–54 Southwark Street
London SE1 1UN

Hardie Grant Books (Australia)
Ground Floor, Building 1
658 Church Street
Melbourne, VIC 3121

hardiegrantbooks.com

British Library Cataloguing-in-Publication Data.
A catalogue record for this book is available from
the British Library.

ISBN: 978-1-78488-457-4
10 9 8 7 6 5 4 3 2 1

Publisher: Kajal Mistry
Editor: Eila Purvis
Cover and Internal Design: Stuart Hardie
Proofreader: Pami Hoggatt
Indexer: Cathy Heath

Colour Reproduction by p2d
Printed and bound in China by Leo Paper Products Ltd.

Reference list:
Curry, A. (2017).
Our 9,000-Year Love Affair With Booze.
National Geographic. [online]
Available at:
nationalgeographic.com/magazine/
article/alcohol-discovery-addiction-
booze-human-culture